Civilizing Argentina

Civilizing Argentina

SCIENCE, MEDICINE,

AND THE MODERN STATE

Julia Rodriguez

The University of

North Carolina Press

Chapel Hill

Set in Cycles and Arepo
by Keystone Typesetting, Inc.

This book was published with the assistance
of the Anniversary Endowment Fund of the
University of North Carolina Press.

The paper in this book meets the guidelines for
permanence and durability of the Committee
on Production Guidelines for Book Longevity
of the Council on Library Resources.

Library of Congress Cataloging-in-
Publication Data

Rodriguez, Julia, 1967–

Civilizing Argentina : science, medicine,
and the modern state / Julia Rodriguez.

p. cm.

Includes bibliographical references and
index.

ISBN-13: 978-0-8078-2997-4 (cloth: alk. paper)

ISBN-10: 0-8078-2997-8 (cloth: alk. paper)

ISBN-13: 978-0-8078-5669-7 (pbk.: alk. paper)

ISBN-10: 0-8078-5669-X (pbk.: alk. paper)

1. Argentina—Civilization—19th century.
2. Argentina—History—1860–1910.
3. Argentina—Civilization—Philosophy.
4. Argentina—Social policy. 5. Science
and state—Argentina—History. 6. Science
and civilization. 7. Eugenics—Argentina—
History. 8. Social control—Argentina—
History. I. Title.

F2847.R639   2006

982'.04—dc22   2005022336

cloth   10 09 08 07 06   5 4 3 2 1
paper   10 09 08 07 06   5 4 3 2 1

For Charlie

WITH LOVE AND GRATITUDE

Unhappy is the land that needs heroes.

—BERTOLT BRECHT, *Galileo Galilei*

Contents

Acknowledgments

In the time it took me to write this book, I have enjoyed support from many quarters. Funding for postdoctoral research and writing came from the National Science Foundation (grant no. 0132878); the American Association for the History of Medicine, who awarded me the 2004 Pressman-Burroughs Wellcome Prize; the David Rockefeller Center for Latin American Studies at Harvard University; the University of New Hampshire Liberal Arts Dean's Office; and the University of New Hampshire Graduate School. I am extremely grateful to all these institutions for providing the crucial financial support without which scholarly work would be impossible.

Many archivists and librarians in the United States and Argentina played a pivotal role in the acquisition of sources. First, thanks to the patient people in interlibrary loan at the University of New Hampshire. I also benefited from the generous advice and guidance of librarians at the New York Public Library, the Library of Congress, and especially at the Harvard University Law Library. In Argentina, the archivists at the Archivo General de la Nación, in both the document and photo archives, were particularly helpful. The late Alfredo Kohn-Loncarica at the University of Buenos Aires medical school provided me with access to valuable materials. The Archivo Vucetich at the Buenos Aires provincial police in La Plata was unusually forthcoming with documents. Special thanks, too, to Lila Caimari, Mariano Plotkin, Osvaldo Barreneche, José Luis LaRocca, and Horacio, Mirta, and María Victoria Pessagno for their hospitality, friendship, and generosity in Buenos Aires.

The initial stages of research and writing were shaped by the advice and input of Nancy Leys Stepan, Herbert S. Klein, Deborah Levenson, David Rosner, Pablo Piccato, and Lila Caimari. At the University of New Hampshire I have benefited from the expertise of my colleagues, especially Frank McCann, Jan Golinski, Eliga Gould, and Lucy Salyer. I would also like to thank the many undergraduate and graduate research assistants who have lent a hand over the years: Alicia Davis, Darcy Arendt, Christopher Benton, and Keri Lewis, all of the University of New Hampshire; Lucas Waltzer of the City University of New York; and Francis Tsang of Columbia University.

Valued colleagues who generously read parts of the revised manuscript over the years include Nancy Leys Stepan, Kristin Ruggiero, Lila Caimari, Mariano Plotkin, Jonathan Ablard, Amy Fairchild, Bill Harris, Janet Polasky, Eliga Gould, Frank McCann, and Jennifer Selwyn. Above all, *Civilizing Argentina* would not be what it is without long conversations about history with Charles Forcey. Special thanks, too, to Jeannette Hopkins for helping me find my voice and for teaching me so much about writing. At the University of North Carolina Press I was fortunate to work with Elaine Maisner, who showed an early interest in the book and provided invaluable advice along the way. I also extend thanks to Bethany Johnson for the excellent editing and to David Hines for his assistance at the press. Thanks, too, to two anonymous readers for their helpful suggestions.

Finally, I am blessed to be surrounded by a community of friends who sustain me during good times and bad: Sven Beckert, Lisa McGirr, Maureen Donnelly, Susan Markens, Leah Cohen, Tina Nadeau, and Karen Matso. Tally Kampen, as always, deserves to be singled out for her love, support, and advice. My parents, Orlando and Phyllis Rodriguez, my grandmother Marta Rodriguez, my sister-in-law and travel companion Elizabeth Soudant, and my in-laws Pete and Linda Forcey supplied comfort, advice, and encouragement.

The greatest appreciation goes to my incredibly patient husband, Charles Forcey, and our children, Kai and Liv. You make my life a pleasure, and this book is for you.

Civilizing Argentina

Introduction

Argentina is a country burdened by its past. In August 2003 more than a hundred people convened at a Buenos Aires city council meeting to argue about whether a three-block strip of Sarmiento Avenue, which honors the modern nation's most illustrious figure, Domingo Faustino Sarmiento, president of the republic from 1868 to 1874, should be renamed after his archrival, the nineteenth-century dictator Juan Manuel de Rosas. A Peronist party official had proposed the street naming as a means of fostering national unity at a time of economic and political crisis. A heroic figure to some, Rosas embodied to others the worst of Argentina's legacy of violence, tyranny, and despotism. "We have been unable to incorporate our past," said Juan Manuel Soaje Pinto, a descendant of Rosas, "and that is why we are still a nation in convulsions."[1]

That seemingly trivial dispute over a street name goes to the heart of the central conflict in Argentine history, a nearly 200-year battle over the definition of "civilization." How does one know when the nation has achieved civilization? What would it mean to civilize Argentina? Which methods would be acceptable to make it so? Should the country resort to violence to create, protect, or enforce civilization?

In Argentina's modern history, periods of relative democracy and civil peace have alternated with authoritarian regimes. In the nineteenth century Rosas and Sarmiento represented two distinctly opposite Argentinas—one rural, traditional, and authoritarian; the other liberal, cosmopolitan, and democratic. Nicolas Shumway, in his book on Argentine intellectual history, characterized the nineteenth-century political intelligentsia, the progenitors of the nation's "guiding fictions," as bearing a "peculiarly divisive mindset. . . . This ideological legacy is in some sense a mythology of exclusion rather than a unifying national ideal, a recipe for divisiveness rather than consensual pluralism."[2] Those elites who thought of themselves as "conservatives," associated with tradition and provincial interests, sought to protect the local social hierarchies and structures of the interior. "Lib-

erals" were united in a loose coalition around modernization and the building of the nation in the image of countries like the United States, which they
identified with the free market and with ideals of constitutional democracy.
Liberals held up models of European values and behavior and sought to
integrate the nation into the international realm.

In 1880, after decades of civil war, and for the next four decades at least,
cosmopolitan liberals, the ideological heirs of Sarmiento, known as the
Generation of 1880, dominated the increasingly centralized nation and attempted to shape its political culture. Advocates of open trade, they sold
Argentina's natural resources to foreigners. Late nineteenth-century Argentina, under their watch, experienced a great economic expansion. It had
seen meteoric development in its export sector after 1870, a dramatic production of wealth, and the expansion of state structures, construction, and
cultural life—at least in the capital city. The booming cattle industry funded
a transformation of Argentine society. Government expanded its offices;
new elegant buildings, sprawling parks, and boulevards sprung up; cafés
and stylish restaurants dotted the city landscape; and a highly visible urban
bourgeoisie frequented the racetrack, opera, and social club. Compared to
other Latin American countries, Argentina was surging ahead. It appeared
that the nation had a chance to reach the levels of prosperity and development of its northern neighbor, the United States.

Ostensibly prodemocracy, late nineteenth-century liberals were elitist
in policy. They existed in tension with an emerging populism and proto-
nationalism that would rise to power in the twentieth century. In the 1940s
the populist president Juan Perón mobilized the working class at the expense of the interests of the landed elite, whom he vilified as the "oligarchy." Later, the military regimes of the 1970s and 1980s would pit the
government against its own people by terrorizing large numbers in the
name of Argentine "civilization."[3] The dream of uniting the fractured nation was as elusive as it had ever been. The opposing forces of conservative
traditionalism had returned as a reactionary political movement with a
vengeance. No peaceful transfer of power from one democratically elected
civilian government to another would occur again until the inauguration of
Carlos Menem as president in 1992.

Argentina's historical trajectory—from prosperity to decline and chaos
in less than half a century—has typically been seen by historians and social
scientists as a "puzzle" and an "enigma."[4] With its temperate climate, fertile land, and proven potential for industry and export success, why has
Argentina failed to live up to its promise? Why has it had more in common

with impoverished and unstable neighboring Latin American nations than with the United States and Canada? Scholars of Argentina's turn-of-the-twentieth-century past have rightly observed its promise, resources, and population on the verge of fantastic development, a phenomenon sociologist Carlos Waisman refers to as Argentina's "reversal of development."[5] In this view, the forty years that began in 1880, the beginning of the peak of the export boom and the Generation of 1880's leadership, had lifted Argentina to affluence. These scholars of the "golden era" ask how such a promising democracy could be overturned by the authoritarian regimes of the twentieth century. They place the reversal in the policies of the 1930s and 1940s, especially in the mistakes of Peronism.[6]

But *Civilizing Argentina* takes the contrary view that, essentially, the "golden era" itself was tarnished. The wealth and accelerated development of the turn of the century was not accompanied by the significant redistribution of income or power. Liberalism did not bring progress, let alone freedom and equality, for all, and it was, in practice, ridden with paradoxes of control and repression, for both men and women.[7] In this light, the period of apparent relative peace and prosperity can be seen not as contrary to but as causal to long-term patterns of conflict and authoritarianism. The principal outcome of the liberal democracy that began at the end of the nineteenth century was, in fact, the maintenance of the status quo of social class hierarchy and the concentration of wealth in an oligarchy.[8] Moreover, the liberal state initiated novel means of limiting personal freedom and controlling political dissent. It began to cultivate new methods of repressive social control, race-based exclusionary practices (as in immigration policy), and even state violence—all, ironically and without seeing any contradiction, in the name of civilization, modernization, and science. The seeds of social decay were planted in Argentina's golden season.

Early twentieth-century contemporaries themselves expressed doubts about Argentina's meteoric rise. In their view, the nation could surge ahead, following the trajectory of unprecedented economic growth and progress begun in the 1870s. Or, it could reverse itself and fall back into what they saw as the stagnation, decay, and savagery of the past. It was an inchoate and fragile prosperity at the turn of the century, characterized by boom and bust, as elites were well aware, and the potential for slippage was acute. Constantly comparing themselves to the wealthier, more "modern" nations in western Europe and North America, they despaired for Argentina's seemingly and frustratingly stymied destiny. They feared that Argentina would not achieve its destined progress and the level of civilization that they believed was necessary and deserved. They correctly assessed the potential for the nation to repeat past patterns of, as they saw it, material and cultural backwardness. They worried that they were destined to pay the price for the dysfunctional structures that had taken hold during 300 years of colonial rule. The Argentine economy was at its strongest, true, but it was still undermined by volatility and fragility.

The Argentine oligarchy's singular goal was to hold on to their power and wealth by whatever means necessary. Turn-of-the-century political leaders hoped to transcend the nation's "backward" past and consolidate power in the right hands—theirs—in order to usher the country into a mythical and idealized future greatness. This idealized future never did appear exactly as they had imagined it, although many of the goals of Argentina's leaders of the early and mid-nineteenth century were fulfilled before 1900: consolidation of political power in the hands of the Buenos Aires elite, most of whose wealth originated in the fertile interior; mass European immigration, promoted to build industry; and an influx of British capital. But, along with those accomplishments came problems similar to those associated with modernity in other countries—class and ethnic conflict above all, along with urban crowding, disease, tensions around changing roles for women, and the rising anonymity of a heterogeneous population.

Elites struggled with the open question of Argentina's national self-image at this critical postcolonial juncture. One of the central fictions of the new Argentine national identity was that of a European-based and unified "race." The descendants of Spanish colonists denigrated the native peoples of Argentina and vestiges of African heritage, while they reflected on the inherent "inferiority" of Afro-Americans in Brazil, Cuba, and the United States. Placing themselves firmly at the European end of the racial hierarchy, they subscribed to theories of Nordic racial superiority and asserted that Argentina belonged in that European pantheon. In the parsing of race typical of their day, turn-of-the-century scientists distinguished between Anglo-Saxon and Latin races. They sought to emulate the Anglo-Saxons and transcend the supposedly inherent tendencies of their own Latinness.

Argentine governments since the 1850s had advocated increased immigration from northern European countries because they believed that such an infusion would bring "healthy" traits to the population and thereby rid Argentina of its inferior Latin, Indian, and African racial character. The elites convinced themselves that they were a European nation, and hence racially superior to the rest of Latin America. (It is a commonly held myth in

Argentina to this day, reflected in a popular joke among other Latin Americans that the typical Argentine is "an Indian who speaks Italian and thinks he is English.") At the turn of the century it appeared that Argentina's entire future and progress would depend on the type and quality of its national population, its modern citizenry, then in formation.

Race, the salient term of its time—and considered a scientific category—provided to Argentina's elite a rationale for conceptions of group superiority and inferiority and justification for control. Steeped in the intellectuals' obsession with eugenics and their conception of higher and lower racial types, modern Argentine progressives saw no contradiction in seeking to import people who could provide rejuvenating powers of "whitening." But the rush of hoped-for Nordic and Anglo-Saxon laborers never came; the vast majority of the immigrants were instead "Latin," that is, of Spanish and Italian origin. Scientific and social elites saw such immigrant groups as bringing disease, crime, and dangerous revolutionary ideas, and they questioned these groups' capacity for loyalty and adherence to a single, Argentine identity. Elites had convinced themselves that a sense of stability and order must be established for progress to occur and that scientific ingenuity and state power must join forces to resist new forms of barbarism.

Science, the liberal elite's badge of legitimacy, was co-opted by the state to assume a central role in expanding state power, in requiring inspections of immigrants and in confining potential troublemakers, and in producing new technologies, such as fingerprinting of dangerous persons. The Argentine state, in alliance with influential men in universities and government institutions, used "modern" methods of control in the name of a benign and enlightened national science and progressive governance. This handpicked group of willing, even eager, experts, no doubt tempted by the confidence of the state—the nation's "social pathologists," as I call them—were rewarded with government positions, state funds, conference and publication subsidies, audiences with the president, and in some cases power and influence beyond their dreams. In return, they offered the authority of science to a state that sought to intervene in and control the lives of ordinary people.

At a time of relative wealth and prosperity, members of the Generation of 1880 and their students, the scientist-bureaucrats, were tapped to head the expanding judiciary, prisons, schools, and administrative bureaucracies.[9] In these leadership roles, they sought to apply the models of theory and practice that they had learned at European universities, especially in France. Scientists with a more radical interpretation of human behavior, for instance, or of the causes of crime and social tumult, were progressively more marginalized with little or no voice in government debates. The regime that resulted was utilitarian, modern, efficient, and orderly, but oligarchic, intrusive, controlling, and punitive.

Argentina's historic struggle for civilization was part of a worldwide moment of ferment around state and nation building and transatlantic issues of immigration, travel, industrialization, labor, and social strife. While Europe was grappling with similar social dynamics, in the Americas, and in Argentina especially, a new awareness of a postcolonial revival emerged. At the end of the nineteenth century, state leaders and intellectuals saw for the first time and in hard numbers the potential for the New World countries to outproduce the Old World and to achieve a civilization greater than that of the European homeland. The question for American nations like Argentina, with enormous productive potential, recently freed from the tyrannies of colonial rule and civil strife, was how to channel this potential. Could they outstrip the great European countries? Was the Old World fading and the New World rising?

As in many other countries, Argentina's national identity rested in large part on the identification and definition of the "other" in its midst. Who that outsider was, and how he or she was defined, changed over the years according to transformations both slow and sudden in the larger social and economic context. Argentina, a community of strangers, needed a unifying force. State leaders perceived a large void in their nation, ready to be filled with the new citizen. Since Argentina had few historical figures and no great past civilization to build on (as in Peru and Mexico), nationalistic state leaders focused on creating one. The nation was the new civilizing project. Conditions were ripe for an elitist makeover of society.

Central to the turn-of-the-century liberal statist science of Argentina was a new view of the body politic.[10] Inspired by the medical triumphs against epidemic disease and the achievements of public health, a rising group of bureaucrats saw what it defined as barbarism as a political disease, a consequence of ethnic and racial inferiority. The model chosen by Argentina's reformist elites in the name of progress and civilization was medical. Social problems like poverty, vagrancy, crime, hysteria, and street violence were defined as illnesses. Symptoms were identified, maladies diagnosed, remedies prescribed, and hygienic systems established to prevent reoccurrence. Physicians and scientists were tapped by the government to diagnose, prescribe, and cure perceived threats to national progress in a comprehensive program to regulate in the national interest both the private and the public lives of the Argentine population. Foreign residents—comprising nearly a third of the nation's population in 1914—were subjected to medical and psychological scrutiny. Even hardworking foreigners, believed to be the majority of newcomers, could slip into antisocial behavior. Their ambitions to participate fully in public life seemed to destabilize the entrenched political hierarchy, as workers rallied for better salaries, democracy, anarchism, and civil rights and as women, fully disfranchised, began to demand participation in the professions and the electorate. State leaders feared, correctly, that all such demands could turn violent or revolutionary.

First came diagnosis. The social pathologists measured skulls, conducted psychological examinations, and divided citizens into categories based on what they regarded as alarming traits, behaviors, moral depravity, and diseases. They received funds to measure the bodies of criminals and the incidence of deviance in the larger society. They investigated the social and sexual habits and living conditions of Buenos Aires's workers, criminals, and homosexuals. They diagnosed vagrancy, crime, and political violence as dangerousness. They seized and confined men and women alike to examine them against their will. Female criminals, prostitutes, lesbians, deficient mothers, and schoolgirls took their places alongside male "degenerates" in the laboratories and hospitals.

After diagnosis logically came a need for prescriptions, for cures, and for proposed hygienic systems to prevent reoccurrence. State scientists concocted targeted fingerprinting programs and imposed standards on men and women in their homes, in public spaces, and in public institutions such as orphanages, hospitals, and prisons. Scientists believed it was their duty to protect the motherhood of new citizens, to restrain medically both male and female sexuality, and to engineer healthy and assimilated children. The goal was to breed the next generation of selected and assimilated citizens while simultaneously purging the nation of "undesirables" and rooting out dangerous individuals by police suppression of political demonstrations and surveillance of political dissenters. Scientifically based laws of social control led to segregation, confinement, or deportation for persons regarded as beyond cure. As in other Atlantic nations of the era, the scientist-government elite sought to engineer individuals, groups, and the social body physically, mentally, and spiritually. In Argentina, however, the search for scientific solutions was fueled by a long-standing and fervent desire for a specific but elusive vision of "civilization." A new scientific

political regimen of social and political purification rose to power, reinforced by police and repressive systems, all in the name of prosperity and
progress.

The state scientists, at once medical and social, with pride and hubris
and a self-perceived purpose of improving and purifying Argentina, tested
the accepted limits of science to identify, heal, and prevent new social
pathologies in the name of breeding and training a modern and civilized
citizenry. But their science was neither objective nor universally positive.
Their methods were shot through with traditional ideas about who was
civilized and who was barbaric. The liberal state saw scientific reform as
saving Argentina by eliminating so-called undesirable and dangerous elements. In a tragedy of irony, it endorsed as civilized a new and more comprehensive form of barbarism. The people of the nation were then forced to
absorb the state's efforts. The story of civilizing Argentina through science
and medicine is the drama of a dream and a vision of true progress imagined
and lost.

PART I *Symptoms*

ONE

Barbarism and the Civilizing Sciences

In America, everything that is not European is barbarian.

—JUAN BAUTISTA ALBERDI

Today, the number of unbalanced [individuals] increases due to the erosion suffered by the race in its exhausting attempts to elevate itself to a civilized level.

—FRANCISCO DE VEYGA

Barbarism in a Young and Fertile Country

Argentina, at the tip of South America, was named for the rich silver deposits that European explorers hoped to find there. Its name, in Spanish, meant land of silver, wealth, and money, and like most outposts of empire, the Spanish settlement in Argentina was about profit. Colonial elites wanted to get the most out of Argentina with the smallest input of resources and energy. Standing in the way of profit were the people colonists found there and, in the later centuries, the people whom elites recruited and lured from Europe as immigrants only to label them savages of another kind, equally difficult to control.

In the fertile River Plate region that became Argentina, the Diaguita, Pampas, Huilliche, Mapuche, and Tehuelche peoples were, for centuries, unconquerable and untamable. Landowners and colonial administrators could not force nomadic "savages" into debt peonage or slave labor or harness their productive forces for tribute, as the conquerors did in Mexico and the Andean viceroyalties.[1] The first Spanish expeditions landed in Argentina in 1516, along the Atlantic coastal region, later named the Rio de la Plata (the River Plate), where they encountered the native Querandí. Amerindians killed the first Spanish sailors to set foot on land. Within twenty years, Buenos Aires had been established by a Spanish aristocrat, Pedro de

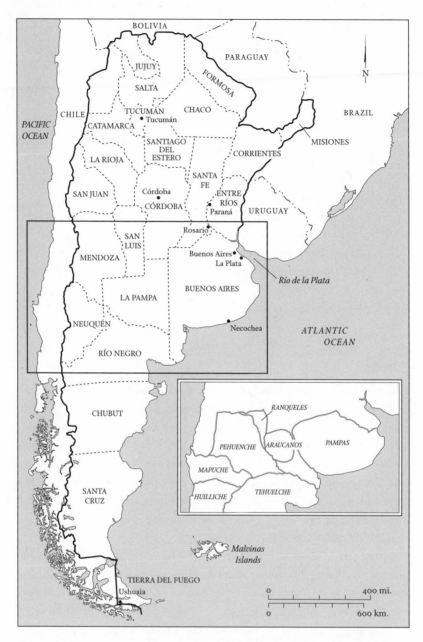

Map 1. Provinces and Territories of Argentina and Native Tribes of the Buenos Aires Region, ca. 1840. (Based on Reginald Lloyd, ed., *Impresiones de la República Argentina en el siglo viente*, 1911, and Donna J. Guy and Thomas E. Sheridan, eds., *Contested Ground: Comparative Frontiers on the Northern and Southern Edges of the Spanish Empire*, 1998)

Mendoza, and from this settlement Spanish troops headed north to explore the territory; however, of the twenty-five Spanish settlements established by the end of the sixteenth century more than half were destroyed by Indian raids. Only about 2,000 Spaniards and about twice that number of mestizos, the mixed-race offspring of natives and Spaniards, remained in a land of numerous tribes of Native Americans.

In 1570 there were 1,000 people altogether in Buenos Aires; by 1660 there were 4,000, most mestizo and Spanish, and a few dozen African slaves. Unlike neighboring Brazil and, to a certain extent, Mexico and Peru, Argentina overall had few slaves, but in eighteenth-century colonial Buenos Aires more than 7,000 inhabitants were black, about a third of the population. By 1887, overwhelmed by the wave of European immigrants, Afro-Argentines were less than 2 percent of residents of Buenos Aires, the urban center of the colonial River Plate and a bustling hub of trade.[2] In 1776 the River Plate was granted autonomy from Peru within the colonial system and named its own viceroyalty, with Buenos Aires as its capital; by 1810 the population had expanded to 42,000. Exports of cattle hides had grown, and colonial goods from other viceroyalties were being shipped out, along with a lively stream of contraband. The first hospital, orphanage, and theaters had gone up, and new schools (for boys only) were accommodating the growing numbers of creole, that is, native-born, Spaniards.

The terms "uncivilized," "savage," and "barbarian" dehumanized and disempowered the conquered and were intended to stigmatize them as inferior, passive, or dangerous. As barbarians they were savages, the dark side of humanity, a threat to the civilized and enlightened, and as such provided a legitimate rationale for conquest, enslavement, and domination. The native Amerindians were seen as a non-European, non-Christian people who must be pacified. The conquerors cited religious and, later, "scientific" texts to back up their claims, equating whiteness of skin and European origin with racial purity and superior culture, habits, and values. After the slave trade to the Americas brought significant numbers of African slaves to Latin America, another meaning of the savage emerged, this time linked to blackness alone.

Argentina's colonial and postcolonial history of race relations is, in some ways, in strong contrast to that of earlier centers of the Spanish empire like Mexico and Peru. In those large and productive hubs of colonial administration and economy, despite similar violence and subjugation of native peoples, a mythical and romantic view of the ancient civilizations arose.[3] Markers and signs of the Incan empire and of native culture, such as monuments, intact Indian villages, and languages, were not destroyed in the onslaught of conquest and survived into the modern period. After the Mexican Revolution ended in 1920, it was official ideology that indigenous and mestizo people, if not the whole nation, were members of a "cosmic race." In Peru, modern national liberation movements adopted the names and images of Incan predecessors, like the Inca leader Tupac Amaru. In Argentina, by contrast, native people were, literally and figuratively, erased from the national memory.

In mid-nineteenth-century Argentina, as in many other nations, the terms "civilization" and "barbarism" were central, organizing principles of the emerging modern state. They galvanized the first leaders of the newly independent United Provinces of the River Plate, as Argentina was known after independence. Domingo Sarmiento, a national intellectual and later president of the republic, in his classic work *Facundo, o civilización i barbarie* (Facundo, or civilization and barbarism), published in 1845, memorialized the leitmotif. Sarmiento saw an epic struggle against Argentina's wild and atavistic forces, represented by the lawless, mixed-race *caudillos*, local strongmen who ruled the interior regions.[4] In the late nineteenth century the federal government required *Civilización i barbarie* for every Argentine schoolchild. Mid-nineteenth-century state policies, from economic measures to population control, were held up to the standard of civilization over barbarism. Elite descendants of the Spanish ruling classes, especially the large landowners and merchants in Buenos Aires's port, sought to remold their land into a place of civilization while, at the same time, envisioning Argentina as an advance over Europe's increasingly crowded and stagnant society. Juan Bautista Alberdi, statesman, diplomat, and author of the Argentine constitution, wrote, "The republics of South America are products and living testimony of the actions of Europe in America. . . . All that is civilized on our soil is European; America itself is a European discovery. . . . In America, everything that is not European is barbarian."[5]

With the growth of the cattle industry—the source of about half of Argentina's exports—a new landed elite had enlarged its power. As trade free of the restrictions of empire became a goal, tensions built between wealthy ranchers and merchants born in Argentina and Spain. In 1806, when the British, engaged in a war that led to Napoleon's takeover of Spain, occupied Buenos Aires, the Spanish militia was no match for the British force, but creole leaders, white descendants of the Spanish-born, secretly gathered troops from the city's population and drove the occupiers out within a few months. Throughout the rest of continental Spanish America, too, native-born Spanish Americans were rising up against foreign rule, using the war in Europe as a catalyst to revolution.

Independence, which brought Spanish rule to an end in 1820, had brought also to an end, in theory, three centuries of barbarism. Nonetheless, military success did not end the civil battle. Argentina, like many other emergent Latin American nations, continued to be rocked by decades of violence and struggle. Conflict between elites, and also between the ruling class and the rural and urban masses, created a fractured society.[6] Argentine traditionalists sought to maintain the basic structures of the colonial period (now in the hands of the creoles), the feudalistic *haciendas* (large plantations that ran on highly exploitative labor and debt peonage). The "liberals"—elites of Argentina's ruling class—looked to foreign models of "progress" in the north Atlantic, especially England and France. They promoted foreign trade, modernized state structures, and new technologies. Provincial elites sought control of local economies, oriented toward the capital Buenos Aires and its port, and sought to turn Argentina outward across the Atlantic, toward the "civilized" world.

Railroads were key to Argentina's growth and prosperity. A railroad system began operation after 1857 with exports from the provinces: wool, then beef and leather. Interior cities, such as Rosario and Córdoba, grew with the new track lines, financed primarily through infusions of British capital and loans. After 1880, the federal government (with the sanction of Congress) approved hundreds of national concessions and the issuing of other types of permits and passes to foreign companies, such as the Central Argentine Railroad and the Buenos Aires Pacific Railroad, both wholly British-owned, to allow the rapid expansion of the rail infrastructure. By 1890 London had invested £157 million, much of it in railroads, and the miles of track went from 6 in 1857 to 5,800 by 1890. While the British owned most of the track, Argentina's government elites considered the railroads indispensable because they enabled imports and exports to rise from 37 million gold pesos in 1861, to 104 million by 1880, to 250 million by 1889, and to nearly a billion pesos by 1915.[7] Tramways, gas, and electricity systems followed. The city began to fence in tracks, to provide guards at street intersections, and, slowly, to build pedestrian overpasses as well. By 1876, with a telegraph link, Buenos Aires had become the nation's communications node with Europe and the rest of the Atlantic world. With railway depots outside the city, well-to-do and middle-class city residents built new homes or summer cottages in the suburbs.

The new elite displayed a growing ambivalence about their infant nation.

Viewing Argentina as empty but fertile, dynamic, and promising, and despite its emulation and imitation of north Atlantic societies, many of the elite felt superior to other countries, especially other Latin American ones. They pointed to their own spectacular achievements of the late nineteenth century, by which time they had eliminated or dominated their rural rivals. The country was growing wealthy from booming agricultural export trade, especially in beef—a process fueled by immigrant labor and the infusion of British capital—inspiring the European saying, "rich as an Argentine." After New York City, Buenos Aires had become the next largest city on the entire Atlantic seaboard. The nation's population skyrocketed from 1.1 million in 1857 to 3.3 million in 1890; most of this growth occurred in the large cities of the littoral or coastal region. Wealthy residents and the federal government financed a succession of beautification projects, including grand buildings, boulevards, parks, and public works. With the help of European capital and German and British architects, Buenos Aires leaders had refashioned a former colonial village into the "Paris of South America."[8]

But Argentina, a young and fertile country only recently freed from colonialism, and with its native people having been exterminated by force, was seen by the elite as empty and in need of a new population. Juan A. Alsina, a public health physician with the immigration service, lamented in 1898, "Of the 2,885,620 square kilometers that make up the surface of the Argentine Republic, there are hardly inhabitants in proportion of 1.40 per kilometer. *That is a vacuum!*"[9] Santiago Vaca-Guzmán, Bolivia's foreign minister and literary figure who lived in Argentina, wrote that same year that the "imbalance between the means of subsistence and the number of inhabitants" in Argentina might provide an outlet for Europe's own "excess of population."[10] By combining its rich natural resources, European tools, and the right kind of immigration, Argentina might exceed other countries, and not only in Latin America, and achieve parity with wealthy industrial countries to the north.

Civilizing the Pampa: Transcending the Nation's Past

In the context of a conflict-ridden and chaotic postcolonial Argentina, elites struggled for dominance of the natural resources in the River Plate region. After 1880, a group of liberal modernizers, politicians, scientists, and intellectuals known as the "Generation of 1880" responded by eliminating the remaining native people, the cowboys of the plains or gauchos, and the rural caudillos, who stuck to the land and their own ways. Military incursions in the 1880s into previously impassable territories in the Pampas region and in the extreme north and south were sometimes accompanied by naturalists and anthropologists seeking data on the inhabitants and cultures of the interior regions. One such study in 1881, by Luis Jorge Fontana, a young government official in the Chaco, a northern region that until then had been an "Indian land," surveyed not only the region's landscape but its native peoples, even as armed forces were wiping them out or driving them to the margins of the nation. President Nicolás Avellaneda credited Fontana's book *El gran Chaco* (The great Chaco) with infusing the nation with a "scientific spirit." The author, Avellaneda wrote, "belongs to the small group of youths who, opening a new way in the intellectual history of our country, have resolved to attempt study and exploration. . . . All these works begin to give a new aspect to our intellectual development."[11] It was a perspective that would come to characterize the governing elite for the next decades, that science was a partner of government and often justified its practices.

Fontana's book included a chapter on the native inhabitants' "intelligence," which was accompanied by a chart of measurements of their heads and feet. Although Fontana challenged the assumptions of "European authors" who "assign a very diminished amount of intellectual faculty" to Indians, he concluded that the Chaco Indians were "more intelligent, cooperative, and above all, more obedient than the Indians of the Pampa and Patagonia"; the southern natives were more bellicose because of their freezing cold environment and thus could "never be as intelligent and able to learn." The Araucans of the Pampa thus exhibited "primitive" behavior.[12] Fontana's stance and tone of racial superiority would pervade much Argentine social and scientific thought for decades to come.

At the same time, a rising young general named Julio Argentino Roca led troops against the natives, considered obstacles to the nation's progress, in one of the most decisive battles of Argentina's history. In just under a year, his forces slaughtered native people, shuttling a few to designated reservations, in what was known as the "conquest of the desert." The land was far from a desert, its pampas the country's richest and most productive farmland. Highly profitable cattle, wheat, and sheep trades would be established there. To Roca, who represented the wealthy elite, it was a desert culturally, void of industry and European settlers, and populated by nomadic and unproductive people.[13]

Also targeted for extermination were the gauchos, who were independent of large landowners and had often sided with the natives or with local strongmen, known throughout Latin America as caudillos. Roughriding cattlemen, many gauchos were of mixed race, a few were Jewish immigrants. The Buenos Aires elite regarded all gauchos and caudillos with contempt. They considered the gauchos to be unmodern and racially and culturally barbaric. Ironically, once the gauchos had been eliminated, they began to be romanticized by poets and writers and, eventually, came to symbolize the new nationalist idea of the "true" Argentine national character (see chapter 9). Today the gaucho, as a legend, is one of the best-known symbols of Argentine culture.

The caudillos, from the small-time local demagogues to regional leaders with armies of their own, a legacy of the decentralized colonial control, were in the eyes of the late nineteenth-century modernizers an even more formidable threat to national progress than the gauchos. The caudillos' influence had been consolidated in many regions by civil strife and anarchy in the decades after Argentina's independence in the early nineteenth century. Their power was derived not from inherited wealth (though some had acquired huge tracts of land) but from their skill in attracting a following through patronage or coercion. Because of their hold over the "rabble," caudillos were difficult to dislodge. The sociologist Carlos Bunge, reflecting widely held views, scorned the caudillos in 1903 as throwbacks for their clan-based feudal societies and their cultlike hold on the people, claiming that "they do not rule by *ideas*, but by *personas* or *their own names*." The local strongman, "magnetic like the eyes of a snake," Bunge said, held his people in his sway.[14]

So, too, liberals of the late nineteenth century considered the notorious caudillo Juan Manuel de Rosas, the Argentine dictator from 1829 to 1852, and smaller-scale strongmen as well, as obstacles to progress and as dangerous elements. In the new forensic psychology that had permeated the intelligentsia, such rural leaders and their followers displayed typically "Argentine" social pathologies that were the consequences of outdated patterns of backwardness, deeply entrenched over the centuries-long colonial period. In the social science studies of national types and character traits, which corresponded to theories of racial hierarchy, the caudillo was associated with Argentina's chaotic, violent, and unpredictable past. "The political organization of a people," as Bunge put it, "is the product of its psychology."[15] José María Ramos Mejía, a prominent physician and public health reformer in Buenos Aires, had condemned figures like Rosas in works such as his influential 1878 book *Las neurosis de los hombres célebres en la historia argentina* (The neurosis of famous men in the history of Argentina) and

Rosas y sus tiempos (Rosas and his times). In 1904 Lucas Ayarragaray, a physician and influential national deputy, published an in-depth study of the historical development of the Argentine national character and its "ethnic and psychological" traits. Ayarragaray, searching for the "psychological origin" of the Argentine nation, found it in rural violence, or "Argentine anarchy." The "decadence" and "indolence" of Spain—common perceptions, and ones that contrasted with images of thrifty, hardworking, and moral northern Atlantic countries—merged with a rural backwardness to "feed our coarse spirit." Ayarragaray wrote, "Good government and the good citizen are not the fruits of chance, nor do they emerge from spontaneous generation from inferior or subverted social states."[16]

Ayarragaray, too, vilified the caudillo and gaucho as rural types. The caudillo, he said, needed the crowd as much as it needed him; the mutual vice and immorality of the crowd and its leader fed each other. The caudillo combined the worst ethnic traits of the Indian and the gaucho, in Ayarragaray's opinion: "In reality, one and the other type blend together in their affinity of fundamental qualities that emerge from a common psychological source. There is a confluence in the spirit of both, a double moral current of ethnic elements constitutive of the race: on the one hand the passivity and duplicity on the native; on the other hand the sullen nature and violent humor of the degenerate gaucho." The result was a "gauchocracy," a "politics of trickery and machinations."[17] Also in 1904, José Ingenieros, Argentina's most famous psychiatrist, published a lengthy essay on "Argentine anarchy and *caudillismo*" in the *Archivos de Psiquiatría, Criminología, y Ciencias Afines* (Archives of psychiatry, criminology, and related sciences), Argentina's first major criminology journal. It was a criticism of Ayarragaray's book and was intended to "scientifically reconstruct" Argentine history. Ingenieros was convinced that Argentine scientists had arrived at the "hour of sociological synthesis" of their nation's past; he took issue with Ayarragaray's lack of analysis of racial factors, complaining that he had not studied the social or economic environment sufficiently. (Ingenieros's disagreement did not prevent his publication eight years later in the same journal of an essay by Ayarragaray on "Argentina's Ethnic Constitution and Its Problems.")[18]

Bunge, in his widely read 1903 book, *Nuestra América* (Our America), claimed to study "the psychology of Spaniards, Indians, and Blacks, taking into account when possible the respective geographic environments in which these races take shape." He described each region's "racial" group, including their "typical," and pathological, traits. In his view, Latin American political formations resulted primarily from racial characteristics. Unremittingly negative, especially, about Latin Americans of African descent, Bunge was equally positive about the prospects for a "whitening" of Argentina: "Today, the census shows that Buenos Aires has an infamously low proportion of blacks. Why this decline? . . . The climate has decimated them, because their lungs badly resist the pampean winds; they have mixed racially and the white race, as the more vigorous one, predominates in those mixtures, . . . and finally, because of the copious influx of European immigration." Each nation in Latin America, though with different mixtures of races and a distinctive national character, had in common "three qualities that support, like immovable columns of rock, the *genius of the race*: LAZINESS, SORROW, AND ARROGANCE." According to Bunge, the "main defects" of the people of the rural interior, their "wiliness and audacity . . . indolence and fickleness," were offset by European immigrants who, once having "adapted and argentinized themselves sufficiently," would improve the national character. They would both "nationalize and creolize" Argentina.[19]

The works of these new social and behavioral scientists all the more reinforced the view that Argentina had, by the end of the nineteenth century, largely left the barbarism of the countryside behind. Victor Pesenti, a lawyer, wrote in a 1901 thesis on the meaning of modernity for civilization in Argentina, "The Argentine Republic has cleared its civilized zone of banditry, the criminal form of struggle between men, which is destined to disappear in its semi-barbaric form with civilization. Today, it is concentrated in the underpopulated regions, like Neuquén."[20] And Ingenieros, although he ascribed the same importance as other writers to the caudillo as a rural scoundrel, if not a violent criminal, believed that due to the evolutionary process of—"to use Sarmiento's exact word"—civilization, the corrupt political culture of *caudillismo* was disappearing. Ingenieros echoed Bunge's sentiment that "slow progress through effort continues, and not coups d'état and demagogic impulses. . . . In a word, Evolution and not Revolution!"[21]

"To Govern Is to Populate": Importing Whiteness

Roca's military slaughter in 1879–80, in the perspective of the Argentine elite, had cleansed the land ethnically and paved the way for an unprecedented economic boom, which the descendants of Spanish landed wealth would spearhead. Argentina could now surge ahead or fall back into the stagnation, decay, and savagery of its past. Liberals in the free market sense, staunch secularists, and converts to Enlightenment thought, Argentina's elite would years later be described derisively by Juan Perón as the "oligarchy" for their wealth, their monopoly on politics, and their resort to rigid social control. But at the end of the nineteenth century these leaders enjoyed a virtual monopoly on Argentine state power and enforced their control through persuasion and force.

Political events of the late nineteenth century were dominated by Julio A. Roca, president of Argentina from 1880 to 1886, and again from 1898 to 1904. Roca had come into office in the wake of his successful military campaign against the Pampas Indians and against, as he saw it, all provincial *caudillaje*: "We have conquered the mentality of the *Montoneros* [guerrillas or insurgents]. We have developed ways to solve problems that appeared beyond us."[22] Once in power, Roca aimed to consolidate and maintain central power of the nation's economy and polity, adhering to the program of material progress laid down by European and North American models. He welcomed foreign capital and European immigration and opened the country's interior to agricultural development. With deep distrust of the popular masses, including the fledgling labor movement in the cities and any signs of direct democracy, Roca limited electoral rights and restricted the process of naturalization and other forms of political participation. Most male citizens were banned from electoral politics until 1912, when secret balloting was mandated; women were not granted the vote until 1947.

The economic boom of the late nineteenth century led to Argentina's stunning transformation from colonial backwater to an urban, industrial society largely based on exploitation of agricultural resources, in particular, cattle and wheat, and aided by foreign investment, mostly British. After 1870, masses of immigrant seasonal labor, some actively recruited from Italy, arrived to toil on the pampa, in the slaughterhouses, in shipyards, and later on, in Buenos Aires's burgeoning factories. After 1880, the gross national product increased by about 6 percent per year, comparable to that of Canada and other industrializing countries. While lagging behind the United States in absolute wealth, Argentina's rate of expansion for most of the twenty years before 1914 was more dramatic, with its per capita income at times equal to that of Germany and Holland and higher than that of Spain, Italy, Sweden, and Switzerland. Growing rates of literacy, life expectancy, and consumer activity, all indices of social progress, confirmed the upward economic trend.[23]

Though the roots of the ruling social class, or "oligarchy," that Roca led were in the great landowning class of the fertile Argentine pampa, the elites of the interior provinces now identified with the cosmopolitan culture of the capital city. Many sons of the powerful Spanish colonial families, who had come as conquerors in the colonial period, intermarried with early English, French, and German immigrants. They thought of themselves as "liberal," and by the standards of the time, they were. They believed in and fostered free trade and rapid capitalist development. They distinguished themselves from the federalist forces of the previous era, who sought independence for the provinces at the expense of a centralizing Buenos Aires and who, to the modernizers, represented stultifying provincialism, Hispanic traditions, and Catholicism. They adhered to progressivist, centralizing, and cosmopolitan views and policies and sought to subordinate the interior provinces to a centralized national apparatus to oversee the nation's economic and cultural advancement. Relying as they did on the models of England and the United States, they also sought, at least on paper, to develop democracy and civil liberties, crafting a constitution that guaranteed freedom on speech and religion and that created a national structure of representative democracy.[24]

Nominally democratic, Argentina's oligarchy of the late nineteenth century fought to retain its monopoly on elected office. Loyal to its ties to powerful economic families, the new leadership fostered the economic interests of foreign capital, great landowners, and wealthy merchants, their "liberal," progress-minded view countered by a strong current of traditional, provincial, and nativist and Hispanist allegiances. Both liberal and conservative elites, however, identified with northern European culture in disparaging the nation's original Indian inhabitants and its few Afro-Argentines, as well as the darker people in neighboring countries. The phrase "to govern is to populate," coined in 1852 by Juan Bautista Alberdi, was invoked to support the Eurocentric elite's conviction that Argentina's new citizens should come from abroad. From the 1860s, the mandate to attract new citizens—above all from desirable, preferably northern, European lands—was part of official government policy. Argentina's ruling government, a coalition of wealthy, elite men with grand goals for their nation, now sought to make the new immigrant masses who had begun to flood into Buenos Aires's port conform to their own goals of progress and national glory. This elite sought to alter the character of the country through sheer will, through an imitation of what they saw in the north Atlantic (including the former British colonies of the United States, with which, in their view, they had much in common), and by importing what they hoped would be a racially and culturally malleable population.

Until the 1880s, Argentina's immigration laws were inclusive with few limits on immigration. The elites and state builders looked to the United States, which had benefited from (predominantly) English and German influxes in the mid-nineteenth century, as a model of a way to build an urban industrial proletariat and to populate the plains and farmlands. Many Argentine leaders, like Domingo Sarmiento, observed critical parallels between the United States and Argentina: vast tracks of "unoccupied" land, largely nomadic and dispersed Indian populations, and a fertile breadbasket region to provide a vibrant agricultural economy. They put their confidence in European immigration to fill the "void" left by the military elimination of barbarism in the Argentine interior.

But Argentina's liberal immigration policies of the late nineteenth century had unexpected and unintended consequences that threatened to make the country a victim of its own success. As the immigrant crowds began to dwarf the rural population, elites discovered an urban pathology. They concluded that immigration and urbanization, the potential salvation for a barbarous nation, had potential perils as well. The nation was overwhelmed by a new foreign ethnic and racial wave, not northern at all, that seemed to bring not a civilizing influence but a new barbarism. The Argentine elite began to discern alarming symptoms of violence, chaos, and disorder among the surging immigrant crowds that swarmed in the nation's urban centers after 1870. Ethnic enclaves, overcrowded and poverty-stricken, were filled with foreigners who spoke in impenetrable dialects and voiced dangerous ideas. Worse, as the century came to a close, turbulent strikes and acts of terror occurred with alarming frequency, including an assassination attempt on President Manuel Quintana in 1905 and a bombing of the opera house in 1910. Fearing that an alien crowd could turn violent or revolutionary, Argentina's leaders began to seek ways to control or expel them.

Francisco de Veyga, a leading forensic psychiatrist, believed Argentina's need to modernize and import large numbers of immigrants had created a whole new set of problems. Witnessing an influx of dangerous types from abroad, especially anarchists, Veyga wrote in 1897, "Today, the number of unbalanced [individuals] increases due to the erosion suffered by the race in its exhausting attempts to elevate itself to a civilized level."[25] Anarchism, considered by many the most explicit challenge to landed, business, and government interests, came to be seen by the Argentine elite as the biggest peril of the 1890s. State leaders found themselves in a bind: they needed immigrants for national development and to populate their country in pursuit of national greatness, but they did not like the ideas the immigrants brought with them. Suppressing such ideas and controlling rising social conflict—crime, riots, strikes, and terrorism—became crucial to national goals of successful development, as well as accelerated profit and growth, and even civilization itself.

Leaders of the oligarchic ruling party, scrambling for strategies to suppress, if not eliminate, social conflict while preserving "healthy" immigration and productive and obedient mass labor, concluded that they could harness modern and scientific techniques of social control by shaping and guiding the nation's foreign population and, when those techniques failed, by confining and excluding those who could not be molded. To this end, leaders encouraged immigrant labor but established mechanisms to control immigrants' behavior and to prevent their ascendance in the social and political hierarchy. In this effort, the most effective strategy they adopted was the withholding of citizenship. A series of residency measures passed by liberal governments after 1860 provided essential business and property rights to immigrants but made political naturalization a long and arduous process. Police officials, the gatekeepers of the system, were known to discourage immigrants from applying for citizenship.[26]

Immigration thus emerged as the principal defining experience of modern Argentina. Between 1871 and 1914, nearly 6 million immigrants arrived, with about half settling permanently.[27] Bunge, son of German immigrants himself, wrote in *Nuestra América* that Argentina's "generous politics, liberal laws, rich production, and eternal blue sky attract and conquer for always the hearts and wills" of settlers.[28] The government established open and generous immigration laws, set up commissions to promote immigration, and funded recruiting missions and advertising campaigns in Europe. Its strategy to protect large landowners turned out, however, to confine the majority of immigrants to the cities. Programs to shuttle new arrivals to the nation's "empty" land and agricultural areas proved dismal failures, with few small plots available for family purchase; and despite the vast tracks of unsettled land, no government plan had been set up to free family plots from the large estates. The percentage of foreign-born residents of Buenos Aires, already a substantial 36 percent in 1855, ballooned to 52 percent by 1895. The vast majority of the immigrants were of "Latin" origin, the majority Italian- or Spanish-born. By 1914, native Italians represented 39 percent of foreign-born Argentine residents, and Spaniards more than 35 percent.[29]

In promoting European immigration, the elite assumed the superiority of the "white race." Argentines' view of race in general employed a continuum of skin color, in which whiteness was associated with progress. The

Culture
race

elite hoped that immigration and intermarriage in Argentina would provide a whitening model to the rest of Latin America.[30] Underlying this view were the centuries-old Hispanic concept of "purity of blood" (*pureza de sangre*) and a belief in the lack of humanity of people of color. "Scientific" theories of racial superiority of northern European "races" did not challenge, but rather rationalized, these antiquated ideas.

The idea of race in Argentina, however, encompassed not skin color alone, but also a European cultural hierarchy in which southern cultures were viewed as intellectually and culturally inferior. Italians, for example, were thought to be stupid, lazy, sneaky, and violent. *En la sangre* (In the blood), an 1885 novel serialized in the newspaper *Sud-América* and written by politician turned literary figure Eugenio Cambaceres, told of an Italian immigrant who becomes a burden on society. Cambaceres, couching his views in naturalistic metaphors, fed on stereotypes of the Italian immigrant as "imbecilic," deceptive, and parasitic.[31] His wildly popular book influenced lawmakers' debates on limiting immigration at the turn of the century. So, too, psychiatrists and criminologists considered Jews (often called "*rusos*," Russians) more likely than other Europeans to suffer race-specific types of psychosis and to engage in anarchism.[32]

Discussions of race reflected a deep identity crisis within Argentina. Members of the established wealthy class, some of them immigrants themselves (and some of them quite recent, from France, Spain, or Italy), took a complex, often contradictory, stance between favoring national autonomy and imitating Europe. They relied on Europe, foremost, for capital and as a market for their exports, and they looked to it for ideas and culture. They emulated European politics, architecture, and art, but they feared its social problems, especially its seemingly rampant urban tensions. They had hoped that new immigrants would resemble the dominant groups that had built the United States in the early to mid-1800s—Dutch, French, English, Scottish, German, and Swedish—but found in their stead southern and, to a lesser extent, eastern Europeans, who were fleeing poverty and oppression to seek fortune in the New World. The Argentina of 1880 had become a polyglot, heterogeneous, and urban society, a troubled society in flux.

The Generation of 1880: Making a Great Nation

To build the nation, President Roca and those who followed him in office relied on collaboration with liberal intellectuals of the Generation of 1880, who had inherited some of their fundamental beliefs, ideas, and prejudices from the Generation of 1837, a postindependence cadre that included Alberdi, author of the 1852 constitution, and Sarmiento. Writing from exile imposed on them for their opposition to dictator Juan Manuel de Rosas, and frustrated in their struggle against the forces of tradition and provincialism, these founding liberals either reviled or sought to improve all that was native to Argentina. The earlier patriots had established the tradition of looking outward to Europe and North America for their models of economics, politics, and culture, adhering to British theories of democracy, law, and free markets. Indeed, Sarmiento concluded his *Civilización i barbarie* with the assertion that the "age of tyrants" would soon be over and that European wisdom would lead Argentina into a new civilized age.[33]

The Generation of 1880, located in the nation's capital, enjoyed the respect and deference of a society enthralled with science, especially its medical breakthroughs and industrial inventiveness. It took no great leap of faith, therefore, to believe that science might diagnose and heal the social ills of modern human society. Enamored with fashionable theories such as Social Darwinism, which stated that some people or races were more fit for survival than others, and positivism, which stressed the scientific verity of observable facts, Argentina's modernizers worshipped science as the only true means to national salvation. Using science's rational methods, they strove to achieve progress by reversing all that their country seemed to represent: backwardness, provincialism, and strict Catholicism. They were united in their confidence that, by relying on science and reason, they could, in the words of one of the most illustrious members of the Generation of 1880, Eduardo Wilde, famous physician and president of the republic in the 1870s, "make a great nation."[34]

If Buenos Aires had been a source of pride for Argentina, however, some of the effects of its accelerated modernity now became a source of shame. Commentators on the city's growth could not ignore the criminals, beggars, prostitutes, angry workers, and outlandish and disheveled immigrants who crowded Buenos Aires's streets. Most alarming, however, were the anonymous, seemingly threatening, perhaps even revolutionary hordes. The volatile economy, with great peaks of prosperity punctuated by crises, as in the United States in this period, caused instability as well. After a devastating financial crash in 1890 that dealt a strong blow to the ruling oligarchy, economic crisis was linked to affronts to oligarchic rule and even to capitalism itself.

In the 1880s many of Buenos Aires's wealthier residents, to escape the port vapors and raucous city crowds, began to move into northern boroughs of Palermo and Belgrano, where they built grand houses and gorgeous parks. At the same time, the government's choices in constructing the new city tramways and other public works conspired to push poorer residents of Buenos Aires southward into barrios like Barracas and Boca. As a consequence, by the 1890s most immigrants were concentrated in tenements, called *conventillos*, in the city's central and southern areas, and these mushrooming, seedy working-class neighborhoods, in turn, jeopardized Buenos Aires's bid to become one of the world's cosmopolitan centers.[35] The elite now saw the teeming tenements and unruly neighborhoods they had created as a blemish and as a breeding ground for what they saw as social pathology, and they began to identify the growing numbers of immigrants with negative by-products of urbanization, like crowding, pauperism, labor unrest, and above all, crime.

It was in this highly charged atmosphere that the nation's prominent physicians, psychiatrists, sociologists, and legal reformers assumed the tasks of diagnosing and curing the myriad social pathologies: dirt, disease, crime, prostitution, vagrancy, and violent class conflict. Scholars, themselves members of the elite or rising professional class, carried into their laboratories and studies and displayed in their reports and books much of the elite's ambivalence about the European influx of ideas and people. Most significant, they shared the conceit that these populations could be quarantined physically and politically from the "true" Argentina, subjected to scientific study, and assimilated or disciplined. With equal efficiency this leadership defined the problems of crime, disease, mental illness, poverty, overcrowding, and public disorder in "scientific" terms and implemented scientific programs to address these issues. They radically expanded the public health infrastructure and built hospitals and sewer systems. They supported public primary education, the expansion of the University of Buenos Aires and the building of universities in the interior, and the establishment and modernization of university science departments, scientific journals, and police laboratories.

An important social base of the scientific elite was the newly consolidated and expanding University of Buenos Aires, no ivory tower but rather a central and powerful cultural and political organ of the capital and nation. Though the university's faculty was relatively small—the medical school had only twenty-three full professors in 1888—its members were more often than not the leaders of their respective professions, particularly in law and medicine, and mentors to the future leaders of the nation. Most of the university's graduating classes were the sons of the nation's economic, social, and cultural elite, who were destined to inherit wealth and power as legislators, judges, and government appointees. Faculty members themselves moved in and out of government service.[36]

These intellectuals and bureaucrats, seeing disorder as a social illness, now adopted a medical model to identify causes and to build an agenda to cure and prevent further eruptions in the national body. They hunted for manifestations of such disease, observed and recorded multiple symptoms, and promoted the ascent to power of scientific professionals who might cure the nation's ills. The views of Victor Pesenti, a noted lawyer, reflected the degree to which the medical model of social problems had been absorbed by the elite. Pesenti wrote in 1901, "Civilization is the most fertile development, the most beautiful of the human energies, but it also possesses an infectious virus of great potency. Along with the splendor of artistic labor and scientific industry, it accumulates gangrenous products: laziness, poverty, madness, crime, suicide, and weakness."[37]

Physicians and state officials, working in tandem, set out to construct a program to ensure that their nation would survive its civilizing process. Their aim was not new—"civilizing" and simultaneously protecting the vested economic and political interests of an elite—but their methods and strategy were. Military campaigns, they seemed to realize, were increasingly unacceptable in the modern age. Seeking new means of managing social problems and engineering their society, they created and fostered a new class of professionals and endowed it with the means and authority to carry out the state's agenda. The Generation of 1880, and those who followed in their footsteps, defined for Argentina a scientific worldview that carried power and legitimacy. With impetus, funding, and encouragement from the oligarchic state, they would shape Argentine political culture for decades to come.

TWO

The Rise of the Social Pathologists

Merging Science and the State

> Intelligence is a vulgar quality of the Argentines. It is converted into talent
> through the labor of pertinent study; it will not pass from sterile frivolity if not
> fertilized by constant work discipline. Today's youth will become legion with
> men of talent in science and letters if they model this rough clay of their native
> intelligence.
>
> —JOSÉ INGENIEROS

As the nineteenth century drew to a close, Argentina emerged as an up-and-coming and promising station in the world of transatlantic ideas. The Argentine state took pains to promote its investments in science by sending ambassadors and reams of published research abroad. Argentine scientific elites looked to Europe for both inspiration and validation. They basked in moments of recognition. The British scientist Sir Francis Galton wrote to Juan Vucetich, a scientist at the Buenos Aires provincial police, to praise Vucetich's fastidious statistical methods. The noted Italian criminologist Cesare Lombroso lauded the Argentine neurologist José María Ramos Mejía as "one of the most potent thinkers and great alienists in the world."[1]

Already known as a source of raw scientific data to be collected and analyzed by experts in the European centers, Argentina as a producer of knowledge was now gaining momentum. In the 1880s naturalist Eduardo Holmberg, a shining star of early Argentine natural science and a descendant of an aristocratic German who had fought with patriots in Argentina's war of independence, had founded the country's first journal devoted to natural science, *La naturalista argentina* (The Argentine naturalist). Now he was director of Argentina's first zoo. With others in the thriving Argentine community of naturalists, Holmberg and the paleontologist Florentino Ameghino shared a passion for South American flora and fauna with illustrious Europeans like Alexander von Humboldt and Charles Darwin. They displayed that passion for the world to see and earned international fame for their studies of nonhuman mammals and the native peoples of Patagonia and the Pampas. At the 1899 Paris world's fair, the Argentine Pavilion earned highest marks from the international jury, and Ameghino won a gold medal for his book, *Contribución al conocimiento de los mamíferos fósiles de la República Argentina* (Contribution to the knowledge of mammalian fossils in the Argentine Republic). Ameghino and his brother Carlos, the sons of Italian immigrants, charted the natural history of Argentina and collected hundreds of specimens and fossils, depositing them not in Europe but in the young nation's first museums.

The first science in Argentina had been colonial in its methods and purposes. Specimens gathered had been sent to Europe for inspection and classification. Diseases were fought with the goal of protecting the European colonists. But now, decades later, the flow of information was in the other direction: scientists focused on the needs of the nascent nation, though they were still dependent largely on European theories and models. Projects in the natural sciences and public health demonstrated a broad engagement with Atlantic empiricism, a commitment to evolutionary theory, and the latest advances in sanitation. Despite their own small number, limited funding, and skeletal scientific infrastructure, Argentina's scientists and physicians, sometimes in collaboration with forward-looking lawyers and bureaucrats, were engaged and entrepreneurial in their pursuit of scientific achievement.[2]

Naturalists in their fields of excellence—geology, botany, and paleontology—such as Ameghino, and public health experts, such as Emilio Coni, Francisco de Veyga, and others, were the heroes of the Generation of 1880, which itself set the course of Argentina's scientific engagement. Yet these older men represented the limits of science's contemporary reach in solving social problems, and they were unable to cure the nation's ills that accompanied the economic and social explosion of the late nineteenth century. To fill that vacuum came new sciences of social pathology—hygiene, psychiatry, and criminology, followed by eugenics and *puericultura* (the science of infant and maternal hygiene)—which emerged from Argentine university laboratories and state health departments to serve as allies of the highly political state builders, the self-styled modern, liberal, and progressive members of the Generation of 1880. Together scientists and political leaders nurtured the growth of the professions to resolve what one scientist called "eruption[s] on the skin of the social body."[3] Social problems were filtered through scientific and medical metaphors in novels and popular literature as well, such as in the famous novels of Eugenio Cambaceres that linked criminality, immigration, and anxieties about family and nation.[4] As the principal symbol of their efforts, social scientists took up the age-old Argentine trope of "barbarism" and proposed that the new sciences confront head on its modern forms.

Intellectual and political elites on both sides of the ocean shared the language of science. Objective, empirical observation, in the words of French philosopher Auguste Comte, was deemed the only reliable means of reflecting and understanding the "positive facts" of nature, including human nature. Especially in the second half of the nineteenth century, thanks to advances in rail and transoceanic travel, the emergence of scientific journals, and sponsorship of practical science in new liberal states, scientists could speak of an "international" community. In some fields rivalries (or "schools") tapped into the surging nationalism of scientists and politicians who sought national recognition and advances from their scientific discoveries. On a deeper level, however, most scientists saw their work as part of a larger stream of discoveries to advance universal knowledge and the greater good.

Empirical observation and analysis attracted the attention of young Argentine scientists in the new disciplines. Just as the naturalists had started local and national museums instead of shipping items back for display to Europe, experts in the human sciences in Buenos Aires hoped to continue engaging with European science even as they built national institutions. Intellectuals, lawmakers, and wealthy men of influence focused on Argentina's need for industrialization, secularization, and modernization, especially in housing, education, and hygiene, conveying the power and legitimacy of the new intellectual developments in Europe. Through close contact with scientists abroad, they communicated in study and travel, correspondence, and meticulous reading of foreign scientific publications. Important as Europe was to the Generation of 1880, Argentines imagined their own independent path to greatness and civilization through science. They were ambivalent about borrowing from Europe, highly aware of their own peripheral and underdeveloped status, and afraid they would not break out of it, yet they held out the promise of a nation-in-formation with the potential to achieve modernity and wealth without the problems of Europe.

Scientists and state officials were mutually dependent for the achievement of their goals. Legislatures supported the development of new scientific methods to heal their ailing social body and invested in practitioners' unprecedented power to control their patients. When the arrival of large numbers of immigrants after 1880 raised the specter of criminals and anarchists escaping detection by hiding in sprawling slums, the Argentine government began to reinforce restrictive mechanisms of social control with the active assistance of scientists. Science and politics were two sides of the same coin.

Argentina, the Idea of Europe, and Racial Implications

In 1903 Argentine psychologist Horacio Piñero, invited to the Sorbonne as a visiting scientist, addressed his audience on the progress of experimental psychology in Argentina and expressed its debt to France. "In the field of science, Sirs," he exclaimed, "we follow the French scientific example called 'the mind of the world.' Intellectually, we are in reality French; we live in the echo of your progress, listening to your lessons in all manifestations of intelligence. And, I should affirm, we owe more to you than all other nations of the world combined, for one simple reason: at the beginning of our adolescence, it was the French who directed our steps, instructed the generations who govern the country today and teach our youth in the schools and universities."[5]

Late nineteenth-century Argentine elites established a tradition of Eurocentrism that implied self-denigration and an idealization of all that was foreign. Intellectuals expressed an indiscriminate admiration for European art, music, architecture, food, and language. All educated cosmopolitan Argentine men were expected to speak, read, and write French, if not multiple European tongues. They worshipped the values that they believed made Europeans and North Americans successful—hard work, thrift, and intelligence. These values, they believed, translated politically into democracy, secularization, and rational states and economically into a humming economy and profit. They envisioned an Argentina that, through emulation, would follow the same path to development as western Europe and North America. They imagined themselves more European than South American.

It was not a real Europe that the Generation of 1880 emulated but rather an *idea* of Europe and what these Argentines considered Europe's most civilized and cultured nations, especially France and Britain. But Europe was, in fact, no paradise. It was suffering from its own social problems: crowding, class tensions, social disruptions, nationalism, and prejudice. Argentines traveling abroad discovered that they were not welcomed in northern European countries as brethren of European descent but grouped with their "darker" Latin American cousins, typed as racially inferior. Even Juan Vucetich, the internationally recognized fingerprinting innovator, suffered humiliations at the hands of British colleagues like Francis Galton, who, because Vucetich's work was published in Spanish, wrote to him in 1897 that he could not "properly read your volume."[6]

New and promising ideas sweeping Europe and North America, such as evolution, positivism, and psychology, were finding a ready home in Argentina. Intellectuals, scientists, and politicians who claimed the authority of science in their attempts to restructure the state and the nation imitated European practitioners of modern, secular science. Not far beneath the surface of Argentine scientists' proclaimed adherence to observation and reasoning was a religious fervor. They embraced the new scientific and medical methods with the enthusiasm of the convert.[7] The reformer-scientists of the Generation of 1880, studying crime and other social problems, followed the rise of empiricism, statistics, classification, and biology in Europe. They adopted the far-reaching European philosophy of positivism—a belief in measurable scientific "fact," above all—as the principal instrument of their "civilizing mission."[8] With a near-worship of French culture, Argentina's social pathologists responded to the methods proposed by the French social thinker Auguste Comte, to his belief that social science was based firmly on "scientific principles" like careful observation and quantitative measure, and to his commitment not only to understand society but to reform it. Comte's motto, "Order and Progress," appealed to members of the new Argentine ruling class. They saw it as a sure fix to the monumental "backwardness" of their country.

Physicians, social scientists, and legal scholars alike filtered social pathology through biological lenses, which led to the rise of medical and anthropometric approaches to social symptoms. Darwinian concepts and Social Darwinism, the idea of the survival of the fittest, influential in nearly all fields both in Europe and America, were in vogue in Argentina from the 1870s on. The Generation of 1880 adhered to theories of Darwin and Herbert Spencer as a means of distancing themselves from earlier generations' conservative and religious thought. The appearance of a Spanish translation of Darwin's *On the Origin of the Species* in 1877 helped popularize evolutionary thought among Argentine intellectuals, and that same year, the Sociedad Científica Argentina (Argentine Scientific Society) named Darwin an honorary member.[9]

But these biological ideas had racial implications. Most Argentine scientists, of diverse national backgrounds, tended to assume that northern

Europeans had natural, inborn traits superior to those of southern Europeans. This presumed biological distinction between groups created a myth of racial hierarchy in Argentina itself, whose people, while predominantly of European descent, were by no means ethnically homogenous or uniform. So, too, Argentine scientists and other elites commonly used racial terms to contrast "civilized" people with "primitives." In 1898, when the medical publication *Semana Médica* (Medical week) reprinted an account of head shrinking among "ferocious tribes" of aborigines in the Philippines, its author concluded, "And to think that such savages are considered to be Spanish citizens by our modern laws!"[10]

In Argentina, ideals of white superiority and the need for racial purity were nearly universally accepted across the political spectrum; they became the "scientific" basis for numerous state projects, such as immigration, crime, and public health control.[11] Progress was defined in terms of reversing or preventing "degeneration," a key concept in late nineteenth-century European thought, one that implied decline from some real or ideal form. By then, scientists liberally used the term to refer to social groups—prostitutes, criminals, the insane, and the poor—considered to hinder national progress and civilization. Armed with the methods of modern science, they believed that they could now identify specific signs and symptoms of such degeneration and devise cures and means of prevention.[12]

Argentines, like other Latin Americans, were ambivalent about their own racial hierarchies and tempered theories of biological causes of degeneration with environmental ones; accordingly, they recommended both biological controls and cultural and educational reforms as "solutions" to racial degeneration.[13] That interpretation of biological theories of race, as not fully genetic but as a consequence, at least in part, of social forces, was typical for turn-of-the-century Latin American liberals. Its neo-Lamarckian, or environmental, approach to race and human social evolution was particularly well suited to Latin American realities; it both explained the degenerative effect of "racial poisons," such as alcohol, and justified the social reform espoused by most liberals. Yet, in discussing the "racial" influences on Argentine crime, criminologist Luis Drago admitted that, despite the predominance of Europeans in Argentina, miscegenation and a history of "barbarism" had resulted in atavism.[14]

When would-be Argentine reformers looked to Europe for models to apply to their own population, they saw a promising new field, criminal anthropology, which combined biology, psychology, sociology, and anthropology to investigate causes of criminal, antisocial, and antinational behavior.[15] Thus France heralded its police scientist Alphonse Bertillon, Austria its criminalist Hans Gross, and Britain, Sir Francis Galton. In Italy, the leading if controversial crime theorist, Cesare Lombroso, led the Continent in approaching crime as a "natural" and inevitable problem but, like illness, preventable if approached scientifically. The new criminology distinguished itself from the so-called classical school of criminology, elaborated by Cesare Beccaria and others in the late eighteenth century, by its emphasis on what they called "scientific" (rather than metaphysical) approaches; on the individual offender and a lack of free will in criminal behavior; on "humane" approaches to criminals in replacing punishment with medically prescribed treatment; and finally on a concern with "social defense," giving social stability a higher claim over individual rights. In Europe the inventors of criminology had already attained both scientific and political status. Lombroso, trained as a physician and professor at the University of Turin, who, together with his students Enrico Ferri and Raffaele Garofalo had created what was known as the "Italian school" of criminology, in 1876 published *Criminal Man*, considered a turning point in approaches to crime.[16] A few years later, Lombroso founded a criminology museum in Turin and created and edited with Ferri the Italian journal *Archivio di Psichiatria e Antropologia Criminale*.

Lombroso's principal—and most controversial—contribution to criminological theory was his idea of "*homo delinquens*," the "born criminal," a type of human being believed to represent a distinct species. Lombroso's *homo delinquens* was a single-sex species: male. His 1893 book, *The Female Offender* (contrast with "criminal man"), written with his son-in-law Guglielmo Ferrero, concluded that women were incapable of committing real or serious crimes, and that any inborn criminality in women was expressed in prostitution.[17] Though Enrico Ferri coined the phrase "born criminal," he is best known for modifying the determinism of his mentor Lombroso's theory by asserting the role of social factors in the shaping of criminal behavior. A lawyer, Ferri took an active part in lobbying for changes in Italian law, for penal reform, and for new social policies intended to alleviate poverty. Ferri believed that improving the quality of life would diminish crime. He played an important role for a time as a socialist politician, but eventually he turned to fascism.

Despite Ferri's modification of Lombroso's theory of crime causation, the primacy of biological factors remained the hallmark of the Italian school. It was Lombroso's view that "the criminal was knowable, measurable and predictable, largely on the basis of cranial, facial and bodily measurements."[18] Lombroso's theory of the born criminal (as described by Ferri) located signs of an individual's criminality in his (and, rarely, in her) physical features. In addition to physical stigmata, Lombroso "found" in criminals sensory and functional peculiarities, "asocial" behavioral manifestations, such as tattooing and the use of slang, and a lack of "moral sense." His ideas reflected a Social Darwinian preoccupation with racial struggle, biological selection, and racial progress or regression. He considered the born criminal an atavistic being, with throwback "primitive" behavior, similar to that of the epileptic and the lunatic. Lombroso's theories had an explicit racial component; he often compared European criminals to the "primitive" Australian and Mongolian races.[19]

New works by Italian criminologists, side by side with French theories, enjoyed growing influence in Argentina. In an article in the *Semana Médica*, José Ingenieros, Argentina's prominent psychiatrist and criminologist, referred to Lombroso's *Criminal Man*, and its methods of identifying different criminal types, as "the most complete study to date on the theme."[20] Some Argentines aligned themselves instead with French critics who faulted Lombroso for ignoring environmental causes of degeneration (the French retained a reductionist and deterministic model of the criminal, and their environmental theories had a medical and biological origin, but they perceived themselves as radically removed from Lombroso's biological theories).[21] In an 1898 review of conference proceedings, Dr. R. Lehmann-Nitsche, director of the Museum of La Plata, argued against the use of Lombrosian "craniology," disputing its contribution to anthropology and the study of racial difference. The following year, however, the same journal reprinted a study of criminal heredity in children (mostly girls) that utilized Lombroso's method of craniological measurements.[22]

References to the influence of Italian positivism can be found as early as 1876 in an article translated from Italian in the first volume of the *Anales de la Sociedad Científica Argentina* (Annals of the Argentine Scientific Society).[23] Most Argentine scientists considered the basic concepts of the Italian school to be accurate and useful. Cornelio Moyano Gacitúa, professor of law at the University of Buenos Aires, reflected this view in his 1899 textbook on criminal law, revealing a typically functionalist attitude toward positivist criminology: "Positivism, despite its errors, has lent service to the science of crimes and of punishment. It has founded perhaps for always the method of observation in this science. It has called attention to the factors of crime with the goal of diminishing or ending it. It has linked criminal science with the natural sciences. Finally, it has suggested the necessity of classifying criminals and rationalizing punishment."[24]

"In the Echo of Your Progress": A Transatlantic Conversation

Communication and collaboration between European and Argentine scholars accelerated in the 1880s when Argentine scientists inserted themselves into European debates through participation in international scientific congresses. Scientific congresses, a critical component of the production and dispersion of scientific knowledge, had burgeoned between 1840 and 1914 to nearly 3,000 such meetings internationally in a number of fields. The primary goal of early criminological congresses was to collect and exchange data, with an emphasis on practical measures of criminal administration.[25] Argentine scientists took home powerful new ideas and approaches from the more mature scientific establishments of Europe. Ingenieros, the psychiatrist, in the middle of a European tour, took an active role at the 1905 International Psychology Congress. Emilio Coni, physician, bureaucrat, and the nation's top hygienist, best known for his institutionalizing efforts in public health and public hygiene, represented Argentina at meetings of the International Congress of Medical Science and the International Congress of Hygiene, where he presented his work, *Progress of Hygiene in the Argentine Republic*, originally published in French.[26]

To Argentine scientists the international congresses presented an important opportunity to gain prestigious backing for their plans for social reform at home. The Argentine delegation to the International Penitentiary Conference in Stockholm in 1878 reported that "the decisions made at this Congress will be of great importance to us since they will not fail to exercise influence on our criminal legislation. . . . The approved resolutions will be like the treatment prescribed by a medical doctor to combat an illness. If we are sicker than other countries, we should try to demonstrate it and indicate the symptoms of our social ills so that we may find a remedy."[27] Increasingly, after the late 1890s, Argentines held their own national and regional scientific meetings in Buenos Aires and other cities.[28] Equally important, they initiated study groups and publications to combat social ills. The social pathologists' precocious attempts to build a national science infrastructure were significant, although, as Horacio Piñero remarked in his speech at the Sorbonne, Argentine scientists would remain "in the echo of your progress" for some years to come.[29]

Most Argentine physicians, like the psychiatrist Ingenieros, enacted the ritual of an extended residency abroad. (Some acquired spouses there, as did Ingenieros; after a second tour he brought his Swiss bride, Eva Rutenberg, back to Buenos Aires with him.) Many scientists' frequent letters and essays home were published in local newspapers and scientific journals. Indeed, only those professionals who could show some degree of success in Europe were accorded membership in the Argentine medical establishment. Argentines followed avidly the prominent theories of British, German, Italian, and later Spanish scientists. French and Italian were considered mandatory languages for Argentine physicians, and Argentine journals routinely published articles in those languages.

Upon departing for an extended European tour in the early 1900s, Ingenieros pictured the considerable Argentine brainpower around him as "rough clay." "Intelligence is a vulgar quality of the Argentines," he wrote. "It is converted into talent through the labor of pertinent study; it will not pass from sterile frivolity if not fertilized by constant work discipline. To-day's youth will become legion with men of talent in science and letters if they model this rough clay of their native intelligence." Ingenieros predicted that the "age of improvisation" would soon be over, and "it will be very difficult to occupy a university chair or exercise public functions without specialized abilities."[30] By this time there was a host of "hardworking" and "intelligent" men in Argentina in medicine, law, psychology, and sociology, ready to develop their raw talent. A well-known judge, Francisco Ramos Mejía, spoke of biology as the "parent soul" of all the "natural sciences of man." In the recent advances in biology, and also in chemistry and physiology, he predicted the rise of "anthropometrical" sciences—craniology, craniometry, ethnology, prehistory, and linguistics. The study of "man," he claimed, rested above all on the development of evolutionary theory and the positivist sciences of the nineteenth century. Anthropology in this sense, as the study of man, was no different from studies of other animals. "Criminal anthropology," Ramos Mejía said, "has as its object the natural study of criminal man, as does zoological anthropology the natural study of men in general."[31]

Judge Ramos Mejía's brother, José María Ramos Mejía, was a central figure in the hygienist-medical circle that dominated Argentina's Generation of 1880. One of the first interpreters of European scientific ideas in Argentine government, he had graduated from medical school in 1879. His doctoral thesis was a clinical study of cerebral trauma, and his early work, *Las neurosis de los hombres célebres* (The neurosis of famous men, 1878), was an influential psychohistorical study of Argentine leaders that followed the latest approaches of European psychology.[32] Francisco de Veyga, a colleague of Ramos Mejía's at the University of Buenos Aires, an army surgeon and hygienist, and a formative influence on the young José Ingenieros, was another early advocate of the professionalization of criminology and legal medicine. Veyga spearheaded the campaign to establish an Institute of Legal Medicine and a Legal Medicine Laboratory. In 1901 he was appointed director of the *sala de observacion de los alienados*, the psychiatric observation room, in the Buenos Aires city police department, and the police agreed to provide a prison infirmary for teaching purposes, further cementing the joint criminological investigations among intellectual and government elites. From his research there, he published articles in *Semana Médica*, the new 1902 criminology journal *Archivos de Psiquiatría, Criminología, y Ciencias Afines* (Archives of psychiatry, criminology, and related sciences), and other journals on political criminals, professional criminals, and sexual "abnormalities," including homosexuality. He was a founding member and treasurer of the Psychology Society in 1909 and a criminologist, the author of an early piece on the criminal tendencies of anarchists.[33]

In 1888 these men and others, aligned with the Italian school of criminal anthropology, formed the Sociedad de Antropología Jurídica (Legal Anthropology Society) in Buenos Aires to discuss Lombroso's theories. It was hailed as the first institution outside of Italy devoted to the study of criminal anthropology, as well as the first in Argentina to lend structure to criminological ideas.[34] Its founding members, Francisco Ramos Mejía, José Nicolas Matienzo, later of the Department of Labor, Norberto Piñero, law professor and member of the Penal Code Commissions of 1891 and 1906, and Rodolfo Rivarola, later director of the *Revista Argentina de Ciencias Politicas* (Argentine review of political sciences), saw themselves as Lombroso's disciples. They cited the "necessity to complete European science with anthropological and sociological data from Argentina and indigenous America" and to "study the personality of the criminal as a base for penal law reform[,] . . . [to] study the character of the criminal, establish the dangerousness and level of responsibility," and to reform the law "according to the principals of the New School."[35] They published *Principios fundamentales de la escuela positiva de derecho penal* (Fundamental principles of the positivist school of criminal law), an 1888 lecture by Francisco Ramos Mejía that was a tribute to positivism and an explication of the "laws" of social evolution. Because of the existence of a "human type dedicated to crime," Ramos Mejía wrote, the state must shape repressive and preventative methods. Helpless to prevent the birth of a "born criminal," the law could nevertheless protect society from him through punishment, segregation, exile, or deportation.[36]

An Alliance of State and Science for the Advancement of Hygiene

Building on the promise and foundation laid by its first generation of social pathologists, Argentine government leaders began to develop local institutions for the diagnosis and treatment of a wide variety of social ills, including madness, crime, and public disorder, and made substantial inroads into Argentine law and state practices. Convinced that hygienic structures were necessary to deal with individual and social maladies, the government sought out the social pathologists as experts for state reform efforts, in the words of one commentator, to "give society precision weapons and a scientific strategy in the secular battle against crime."[37]

Reformers first embarked upon building the nation's medical and hygienic infrastructure. Emilio Coni, author of *The Progress of Hygiene*, urged all municipal and provincial agencies to compile statistics and establish a medical police corps to oversee inspections and vaccinations. A national "medical corps" had been set up in 1852, composed of the faculty of the University of Buenos Aires medical school, the Hygiene Council, and the Academy of Medicine. After 1867, the Hygiene Council, with responsibilities for sanitary controls of the city and port and the regulation of pharmaceuticals, acquired the power to impose fines and penalties for infractions against the code of hygiene.[38] In the wake of the devastating yellow fever and cholera epidemics of the 1860s and 1870s, the public health apparatus assumed greater responsibilities and took on more extensive public works. A new "special health corps," all of whose members were required to be Argentine citizens, had been assigned as early as the 1870s to police Argentina's ports and inspect foreign vessels, especially passenger ships. The corps was trained in medical geography, naval hygiene, and "exotic contagious diseases."[39]

Powerful politicians who had been trained in medicine, like President Eduardo Wilde, symbolized the close alliance of medical and state interests. In 1875 Wilde became a member of the Departments of Hygiene and Legal Medicine at the two-year-old University of Buenos Aires, joining forces with Guillermo Rawson, a physician from a prominent Buenos Aires family who had taught the first courses in public sanitation. By 1880, Wilde had entered government hygiene programs as president of the city waterworks and sanitation commissions. Carlos Pellegrini, a lawyer who took the presidency in 1890, nicknamed "El Gringo" due to his maternal English ancestry, was also a public health physician. The heart of this state-science alliance was the University of Buenos Aires, in particular the *facultad de medicina* or medical school. From its example, universities and schools in the interior drew their inspiration, most notable among them the University of Paraná, later known for its advances in scientific method.

Physicians had organized their own first professional group, the Buenos Aires Medical Association, in the 1860s. Psychiatrist José María Ramos Mejía, scion of a prominent family populated with top physicians, judges, and politicians, founded the Argentine Medical Circle (Círculo Médico Argentino) in 1875, a professional group that Coni described as intended for "the study of everything relating to hygiene and national medicine."[40] In 1891 the Argentine Medical Association was formed, with sections on anatomical pathology; nervous diseases and legal medicine; and hygiene, demography, statistics, sanitation (*policía sanitaria*), epidemiology (grouped with climatology, hydrology, and meteorology). The first Argentine professional medical publication, the *Revista Médico-Quirurgica* (Medical-surgical review) appeared in 1865; the more widely read and influential *Semana Médica* was established in 1894. The investment in professional scientific and medical journals, research institutes, and government research laboratories would bear full fruit after 1900, but in the earlier years legal and medical experts acted as public servants and amateur criminologists by attempting to counter disease of the human body and of the social body and by integrating European criminological concepts with implementation to solve growing Argentine problems of crime and social disorder.

José María Ramos Mejía, the author of *Las neurosis de hombres célebres*, *Las multitudes argentinas* (Argentine crowds), and numerous books and articles on Argentina's social pathologies who was most effective in various government posts, undertook a long-term effort to rationalize and modernize hospitals and mental asylums.[41] Ramos Mejía was president of the Public Assistance Office in 1882, and he led the Department of Hygiene between 1892 and 1898 (followed in this role by Eduardo Wilde). The Department of Hygiene, established in 1880, replaced and expanded the Public Hygiene Council. Ramos Mejía later served as president of the Consejo Nacional de Educación (National Education Council) and was a congressional representative in the late 1880s and early 1890s. At the University of Buenos Aires Ramos Mejía had a hand in the founding of the Nervous Pathology Laboratory in 1891. In a letter to the dean of the medical school, he listed his personal donations to the laboratory, including "instruments of psychometrics," vivisection, electricity, and "anthropology," by which he meant a human skull.[42] He founded the Department of Neuropathology in 1887 and a few years later taught and mentored José Ingenieros, the first major figure in Argentina in the formal discipline of criminology.

Ramos Mejía's colleague Francisco de Veyga convinced university administrators in 1899 to establish a Legal Medicine Laboratory, which Veyga envisioned as a research and training ground for medical students. He proposed a joint venture to be run by the medical school, the police, and the city morgue. The government would fund most of the budget, and the university supplied space for laboratory work and classes in legal medicine. Students would examine the mentally ill and criminals with electrodiagnostics, psychodiagnostics, and craniometry. Veyga justified his new laboratory on the grounds of national interest and modernization of Argentina. He also cited the importance of psychopathology and criminal anthropology to future students of legal medicine. Veyga had argued that many instruments needed for the laboratory had not yet been introduced in Argentina, and he took that lack as a troubling sign of the nation's scientific backwardness. "My didactic program is," he wrote, "through national legal medicine, to shine the light of those medicolegal doctors on the sciences that they are distinguished to serve."[43]

The work of public health physicians, known as *higienistas*, now began to gain in importance, as the state committed expanding resources to public health campaigns against epidemics, which in turn provided a model for similar public-private collaboration in the sphere of social hygiene. In 1887 Emilio Coni lauded the Department of Hygiene as critical to the "considerable progress" of sanitary organization in Argentina, in its regulating medical and pharmaceutical practices, performing sanitary inspections, and advising local governments of "public health lacunae." Coni published widely in *Semana Médica* and the *Revista Médico-Quirurgica*, for many years an editor of the former and a director of the latter. He was also director of the *Bulletin Mensuel de Demographie*, and his many public health surveys of Argentine provinces and cities included tabulations of prison populations, hospital services, legitimate and illegitimate births, prostitution, and venereal disease.[44] In 1883, with José María Ramos Mejía, Coni founded the Asistencia Pública, or Welfare Department, based on the French model, to centralize, organize, and regulate the city's hospitals and asylums. As a public health crusader and advocate for social welfare programs, Coni focused on poverty and the living conditions of the poor. His concern with venereal disease, alcoholism, insanity, and other social "ills" brought him into close contact with colleagues like Veyga and Ramos Mejía who specialized in crime, and he similarly advocated direct government intervention in "sick populations," including inspection of living quarters and mandatory vaccinations. He argued that limiting the movement of prostitutes would lower rates of venereal disease.

Hygienists and state officials together were seeking to bring awareness of medical advances to the public. In 1885 municipal officials oversaw the opening of the Scientific Anatomical-Pathological Museum (Museo Científico Anatomo-patholológico) in the National Theater. In an 1891 letter to the interior minister, an official at the Department of Health said that the museum would provide a valuable service to medical students and other professionals and "stimulate and facilitate the development of the study of hygiene, one of the most useful branches of medical science." Modeled on the Parkes Museum in London and Robert Koch's museum in Berlin, the museum was to contain thirteen sections, including industrial hygiene, worker housing, disinfection, dress, infant and school hygiene, gymnastics, food, and "hospitals, poorhouses, asylums, and jails." Emulation of Europe was explicit in the proposal: "The example of more advanced countries who concern themselves first and foremost with health and life, which is the best good and the best force of the individual and the community, clearly demonstrates that much of the scientific evolution of hygiene in whichever country . . . requires scientific concentration and coordination from all branches of study as a point of departure and as an indispensable condition of its development."[45] A city inspector sent to the museum to assess the exhibits approvingly noted the display of "a great number of illnesses and natural phenomena[,] the majority of which cases occurred in European hospitals." He also noted one "unsuitable" aspect of the exhibit—that of genital diseases—and recommended that these materials be set aside in a private room for viewing by men only.[46]

Above all, Argentine hygienists hoped to impose a new medical meaning of abnormality to replace the old moral ones, projecting a collective preoccupation with anomalous biological, medical, and social phenomena. They staged anatomical exhibitions not only to train medical students but also to showcase "anomalies" and "social diseases" to citizens, and thereby establish what was normal and what was not. Experts believed that an educated and concerned public would more readily tolerate newly rationalized and scientific state intervention in their lives.

The Medical Policing of the "Born Criminal"

When medicine turned to questions of social hygiene as a consequence of behavior, it tried to solve problems out of the reach of science. Physicians constructed moral categories of deviance and behavior to explain prostitution, alcoholism, suicide, and crime, paving the way for normative judgments and hence for the medical policing of society. In 1884 scientist Samuel Gache appealed to an early professional association to fight the social scourge of suicide as a moral and a medical problem. Gache saw suicide, like crime itself, as an indicator of "degeneration" and the person who would commit suicide as a "sick organism": "The Argentine Medical Circle, which today is our scientific leader, should bring to discussion these transcendental questions, intimately linked to the character of its institution. . . . The Argentine Medical Circle will put its influence in this struggle for public morality that has as its object to cure an ulcer—the first manifestation produced by a sick organism." Gache, in the *Anales del Círculo Médico Argentino* in 1884, wrote of suicide as a form of neurosis, caused by the condition of society, like crime. It was a direct and inevitable consequence of modernity or the "level of civilization" of a given country. Citing statistical data on suicides in Buenos Aires for the years 1881 through 1884, supplied by police forensic pathologists, Gache compared Argentina's suicide rates to those of several European countries and, to bolster his case, cited numerous experts, most of them French. He concluded that, like crime rates and as in Europe, suicide was on the rise in Buenos Aires.[47]

If certain medical conditions were now a crime, crime itself was seen as a disease. With the nation's medical infrastructure in place, at least in the capital city of Buenos Aires, physicians turned to the study of crime as a contagious blight. As one criminologist put it, "Crimes are, in a certain way, an eruption on the skin of the social body, sometimes an indication of a grave illness."[48] Miguel A. Lancelotti, a left-wing lawyer and criminologist, wrote in 1898 in the newspaper *Revista Nacional* (National review) that "robbery, suicide, homicide, and the whole sad cohort of crimes fatally follow the law of transmission."[49] A hallmark of the medical approach in this formative period was an expansion of "crime" as a category of diagnosis. It was used as such both literally and euphemistically to mask and justify state projects to control social pathologies, such as madness, vagrancy, alcoholism, and even communicable disease, all of which were to be studied under the rubric of criminal anthropology.

The battle against social pathologies began in university lecture halls and continued by recruiting and training students to be the new leaders for reform and control of state institutions. Courses on crime began to appear in related departments at the University of Buenos Aires. In 1878 Miguel Cané, then a professor in the university's School of Philosophy and Letters, wrote to his dean that "For the first time, there is talk . . . of anthropology. . . . I have initiated the study of the human races . . . in scientific research."[50] At the medical school Francisco de Veyga offered in 1897 the first course in criminal anthropology. Norberto Piñero, in his own 1887 course at the law school, inculcated in his students the importance of following new scientific methods. The course, which was viewed by his colleagues as pathbreaking, was divided into lectures on "the positivist school," crime, criminals, causes of crime, prevention, and punishment, with the goal of determining "the dangerousness of the offender as criteria for establishing necessary repression." Piñero's new approach to the study of crime included "the practical usefulness of classification and identification of offenders." He lectured, too, on the value of statistical data to studies of crime.[51]

With the growing attention to criminology, a number of doctoral dissertations in medicine and law now concentrated on crime and social pathology. Beginning with Pedro Alacer's medical thesis "Madness and Crime" in 1883, some two dozen such dissertations appeared (in Buenos Aires alone) before the turn of the century. Popular university lectures on criminal anthropology and legal medicine found an eager market for publication of pamphlets or books. In law, students were drawn to medical studies of crime and degeneration, specifically on the value of positivist biological and medical approaches for Argentine legislative reforms, with theses like "The Positivist School in Criminal Law" and "The Classification of Delinquents."[52]

Not until 1898 was there a specific journal devoted to criminology nor, for that matter, was there a university chair in criminology. But crime was of such concern to modernizing officials and intellectuals in the 1880s that even without professional infrastructure, numerous full manuscripts and growing numbers of articles in a variety of medical, legal, and general interest journals appeared. A wide array of medical journals served as a conduit for European ideas about crime, mental illness, prostitution, and other social pathologies. Correspondents reported on the proceedings at international scientific and criminological conferences. Medicolegist Francisco de Veyga and criminologist and University of Buenos Aires psychiatry professor Benjamin Solari, editors of the *Semana Médica* in the late 1890s, published articles and legal cases based on the latest European research. Book-length studies on crime, vice, and social disease had appeared in the

1880s, reflecting a nearly wholesale adoption of French and Italian theory in the attempt to modernize—to make more "scientific"—efforts to understand and control a range of problematic human behavior, from begging to murder.[53]

One of the earliest published studies of crime by an Argentine author was the 1888 *Hombres de presa* (Men of prey) by Luis Drago, a legal scholar and legislator.[54] Essentially a favorable engagement with the biological theories of Lombroso, the book also took account of Herbert Spencer, Enrico Ferri, and rival French thinkers. Drago had spent a year as editor of *La Nación*, one of the capital's largest newspapers, in the late 1870s, and from 1882 until his death in 1921 he served as provincial representative, civil judge, criminal judge, and, under President Julio A. Roca, as minister of foreign relations. Argentines were unspeakably proud of the fact that the 1890 Italian translation of *Hombres de presa* was introduced by Lombroso.[55] Drago primarily discussed the application of European ideas in Argentina, but he took pains to point out the local, albeit "secondary," scientific work underway in his own country. He had devoted large sections of his manuscript to discussions of degeneration and the medical conditions of offenders. He praised anthropometry as one of the most important methods of explaining crime and called for a more efficient system of identification to reduce recidivism. His book included an extensive appendix on comparative craniometry.[56]

Other legal scholars joined hygienist colleagues in viewing crime as a medical problem. Rodolfo Rivarola, a professor of law at University of Buenos Aires, worked on the Argentine legal code revisions of 1890 and 1906. He was a founding member of the Sociedad de Antropología Jurídica (Legal Anthropology Society) in 1887 and the Argentine Psychology Society in 1909, and later he published widely on psychology and criminology in the *Archivos de Psiquiatría, Criminología, y Ciencias Afines* and other journals. In 1910 he wrote his renowned *Derecho penal argentino* (Argentine penal law) and founded the *Revista Argentina de Ciencias Politicas*.[57] Norberto Piñero, the legal scholar and professor, was active in criminal code reform, published widely on criminal law, and also held cabinet-level appointments under Presidents José Figueroa Alcorta and Roque Sáenz Peña in the early 1900s. Antonio Dellepiane, professor of law at the University of Buenos Aires and a member of the city commission on jails and houses of correction, published two of the first manuscript-length studies of criminal psychology, *Derecho penal. Las causas del delito.* (Criminal law. The causes of crime, 1892), and *El idioma del delito* (The language of crime, 1894). Like Drago he was a founding member of the Argentine Psychology Society.[58]

These Argentine authors were, for the first time, conscious of the national importance of their own work. While wedded to and dependent on the work of men from other countries, these first Argentine criminologists stated the necessity of attacking local problems. In 1898, with the help of statistician Miguel Lancelotti and the medical student José Ingenieros, Pietro Gori, an Italian journalist with anarchist sympathies, established the journal *Criminalogía Moderna* (Modern criminalogy; the editors explained that they had intentionally spelled the journal's title "criminalogy," instead of the more commonly used "criminology," to reflect the new science's emphasis on the individual offender). Gori conceived of *Criminalogía Moderna* as a European-American collaboration; contributing editors and writers included nearly every recognized criminal lawyer, medicolegist, and criminologist in Argentina and a number of European scientists, notably Lombroso and Ferri. The journal blamed capitalism, not personal physical characteristics, for crime and pointed out that most victims of crime were poor and working-class people. But the journal's main goal was to provide a platform for Argentine scholars to publish and share their research and analysis, its national focus exemplified in analysis of specific local criminal cases, analysis of statistics from the police of Buenos Aires, and discussions of the need for legal reform in Argentina.[59]

That a left-wing criminology journal with links to Italian anarchism could thrive, even for a short period of time, indicates the open-ended character of Argentine criminal science before 1900. However, Gori ultimately could not keep *Criminalogía Moderna* alive because he lacked funds, and because his radical critique of society (in particular, of capitalism) as the cause of crime did not fit with the government's worldview. Since the journal could not survive on individual subscriptions, it folded. Two years after its demise, Ingenieros received funding and institutional support from the government for his own journal *Archivos de Psiquiatría, Criminología, y Ciencias Afines*. Ingenieros by now distanced himself from his youthful socialism and offered a distinctly less radical critique than Gori. Instead of citing workers' inequality and the arbitrary cruelty of capitalism, Ingenieros pointed to specific individual causes of crime and mental illness, biological and environmental.[60]

As the century drew to a close, it became increasingly clear that ad hoc efforts to apply European criminological ideas to existing structures would not be enough for Argentine modernizers. Observing the more developed, seemingly modern structures of Europe, Argentine scientists believed they knew the course to follow: to keep up they needed a modern, professional organization to structure their efforts to fight the wave of crime that seemed to damage their society. Moreover, they needed a strong patron, funding, and infrastructure. State sponsorship effectively marked the transition from private initiatives to government intervention in studies of society's problems. Drago, speaking to his fellow reformers in 1888, urged them to seize the moment. He declared, "It appears the moment has arrived when Argentine publicists and legislators grasp the need to face the serious problems of crime that compromise the very roots of the social order."[61] To defend society, they would need new laws calibrated to the nation's problems. State builders and legal reformers were convinced that Argentina was ripe for such development. As one commentator noted, "of the various Spanish-American states, [Argentina] stands out in penal reform."[62] Argentina, as another stated in a poetic temper, with its "historic virginity appears fit to be fertilized" with the "seeds" of modern change.[63] They began to praise the excellence and regional superiority of their institutions of scientific criminology, such as the Criminology Institute and fingerprint laboratories in Buenos Aires.

Juan Vucetich, a Yugoslavian-born police statistician in La Plata, capital of Buenos Aires province, was one of the most effective at transforming state goals with new empirical methods. Without a formal education or close ties with criminological thinkers emerging in the university, he nonetheless rose rapidly in the ranks of the provincial police's scientific wing, directing the monthly statistical bulletin that reported arrest and crime statistics for La Plata.[64] Two years later he advanced to the head position in the Office of Identification, where he developed his famous method of dactyloscopy, the analysis of individual fingerprints as a means of identification. The nation's most active state scientists recognized the potential of Vucetich's work. Between 1902 and 1914, the *Archivos* published about a dozen major articles on his methods and numerous book and article reviews on the subject. In 1901 the lawyer and criminologist Miguel Lancelotti wrote Vucetich to commend him on his "brilliant lecture" on dactyloscopy delivered at the public library in Buenos Aires. He praised Vucetich's system, stating, "I believe firmly that in the matter of criminal identification there is nothing better that can surpass it, not for simplicity, nor efficiency." Vucetich's lecture was an example of the loftiest scientific ideals. In his letter Lancelotti decried the "public powers [that] still do not recognize the importance that you deserve, dealing with a masterpiece that not only honors its creator and the country . . . but also the very services lent to criminal science and even more so to the public order."[65] Ingenieros, writing in his role as the president of the Argentine Pro-Lombroso Committee, asked Vucetich in 1911 to join the Argentine branch of this international committee, dedicated to building a monument in Rome memorializing Lombroso.[66]

Dactyloscopy, recognized after 1892 as a regional innovation, was employed in most South American nations. The organizers of the South American Police Convention established in its constitution the types of "persons considered dangerous," including criminals and their accomplices; those who "share a common life with habitual criminals," including "foreigners or citizens" who committed crimes abroad; those profiting from the "white slave trade"; and labor agitators, "disturbing the right to work [*libertad del trabajo*] with acts of violence or force, or attacking property." The conference set up by the constitution recommended fingerprinting to identify such "dangerous persons." Its uniform guidelines for reporting information on criminal individuals once they had been apprehended also recommended providing a morphological description, "according to the system of the Province of Buenos Aires," photographs, including frontal and side view facial images, and detailed notes on particularities such as scars and tattoos.[67] Police chiefs in five major South American cities endorsed the guidelines. In 1903 Felix Pacheco, a Brazilian police official, compared Vucetich's science to its European competitors and, especially, to its main Continental rival, Bertillonage. Pacheco wrote in the *Archivos*, "Identifying with the spiritual life of the European world, accepting without discussion all that comes here from there, and in particular, the eternal tributaries of French culture, we never imagined that there could exist something better in this semibarbaric and dark Latin America."[68] One Brazilian medicolegist proclaimed, "No one has a different opinion from mine: the Argentine system surpassed the French."[69]

From Europe Federico Olóriz Aguilera of the University of Madrid wrote in 1910 that, for Spain, the Vucetich method was the most preferable human identification system. Olóriz suggested adding physical descriptions and photographs of subjects to the fingerprint files. Going even further, in his official capacity as professor at the police school, he had revised and modified the prisoner identity card. Like Vucetich, he sought to promote a "National Identity Archive, rigorously classified," in which record cards with physical descriptions and fingerprints would be recorded for every citizen.

Such an archive would allow the state to "apply the advantages of scientific identification to all the acts of social life."[70] The master criminologist Lombroso wrote in an 1893 letter to Vucetich, "I have received your work and frankly I admire it very much. . . . Your innovations [in anthropometry] appear to me to be of great importance." He asked Vucetich to send him a number of samples of fingerprints, "accompanied by descriptions of every one of the criminals to whom they belong." Three years later the Italian master added in a letter, "At least twenty-five years will pass before we in Italy will arrive to where you are now. . . . I have discussed [your] discovery more widely in the new edition of *Criminal Man*. If you consult this book you will see how much I am indebted to your work."[71]

With active state participation, there had been a change in Argentina from ad hoc science, imitative of Europe, to a relationship of symbiosis between elite intellectuals, physicians, hygienists, and state officials. Each strengthened the others. Governments used scientific methods as a legitimated means to reinforce its apparatus of hygienic social control and to begin medical policing of its population. Scientists gained prestige and state support for their work, convinced that they could best fulfill their professional goals through an alliance with the state. The members of the Generation of 1880 had laid the foundation by building medical and hygienic infrastructure, but it was their students who benefited most from the intense interest and involvement of the state.

The state stepped in as the patron of Argentine scientists, selecting and co-opting them to serve its own goals. José Ingenieros had said optimistically in 1902, "The march of civilization brings new crimes, but also many means of avoiding them."[72] Men like Ingenieros, an energetic psychiatrist and criminologist, and the police scientist Juan Vucetich embodied that promise, and to the state they seemed a worthy investment. Together, scientists and state officials crafted a program to rid society of its human problems. They built their program around the concept of the "born criminal," which they believed posed a myriad of dangers to the nation's health and progress. Their efforts were a logical outgrowth of the blending of late nineteenth-century public health and sanitation efforts with a moralistic agenda of social and political control, designed to counter what seemed to them to be the threat of dangerous and abnormal individuals and rising, out-of-control masses.

PART II *Diagnosis*

THREE

A National Science to Investigate the "Abnormal Individual"

> We aspire to create a national science, a national art, a national politics, a
> national sentiment, adapting the characteristics of our multiple originating
> races to the framework of our physical and sociological environment. Just as all
> men aspire to be someone in their family . . . we also aspire that our nation will
> be one with humanity.
>
> —JOSÉ INGENIEROS

In 1905 President Julio Argentino Roca invited the psychiatrist José Ingenieros to accompany him on a diplomatic mission to Europe. It became a
two-year tour during which the young scientist pondered Argentina's place
in the "civilized world." American lands like Argentina might become great
empires, he wrote home in an effusive letter from Berlin: "Germanic and
Anglo-Saxon groups—England yesterday, Germany today, and the United
States tomorrow—arrive at their moment. Their historical role, through
fecund and intense action, is equal to that of the great empires that fill a
chapter of the human chronicle." "Greatness," he explained, was a consequence of national and racial types: "The average type of German, English,
or *yanqui* man possesses common psychological traits that belong to the
collective imperialist sentiment." Although Ingenieros singled out northern
countries, "there are no sociological reasons to believe that the European
continent will eternally hold the first place in human civilization." Might
not nations like Argentina and Australia, "orienting themselves toward new
ideas, ceaselessly renewed[,] . . . acquire a cardinal influence in the civilization of the entire world?"[1]

Catching up with and even surpassing Europe would depend on Argentina's reversing the symptoms of barbarism, degeneration, and abnormality, even of racial inferiority. Ingenieros arranged racial and national groups

in a hierarchy of "civilization." Yet traits could be altered. This contradictory view, inborn character versus environmental effects, reflected his and his fellow social pathologists' continuing ambivalence about the Argentine people. Empirical investigation, including thorough and invasive examinations of bodies, minds, and heredity, was a path to understanding the causes and hence possible solutions. The poor, mostly immigrants, of Buenos Aires and women of all classes were to be subjected to medical examination and treatments. The work of the social pathologists started from these assumptions, which led to empirical studies, and which in turn supported their a posteriori conclusions about deviant types and pointed to social programs to cure, control, or prevent pathology.

If Argentine scientists never fully rejected their European models, in the first decade of the twentieth century they began to promote a nationalist vision of their work. In this, they followed the course led several decades earlier by members of the intellectual cadre known as the Generation of 1880, such as José María Ramos Mejía and Luis María Drago, whose nineteenth-century texts first addressed the nation's growing social problems. Especially after 1902, criminologists could draw on rising state interest and national investment in new criminal justice structures devoted to the study of social pathology: sponsorship of research, close cooperation with legal and penal institutions, and publishing opportunities. Their common task was to build a "national science" to heal or prevent national sickness and to carry out a plan for Argentine achievement. As Ingenieros wrote on returning to Buenos Aires in 1907 from his long tour, "We aspire to create a national science, a national art, a national politics, a national sentiment, adapting the characteristics of our multiple originating races to the framework of our physical and sociological environment. Just as all men aspire to be someone in their family . . . we also aspire that our nation will be one with humanity."[2] Society was their laboratory. But no quantitative project, however precise, is bias-free. What the scientists chose to measure was itself selective. They measured only certain types of crime, and certain people's bodies. And their interpretation of their empirical data was subjective.

"A Study of Our Own Criminality": Measuring Social Pathology

By the 1890s Buenos Aires scientists were known for their extensive and reliable crime statistics. Not long after the first national census of 1869, detailed statistics on crime, suicide, fires, and the incidence of mental disease were compiled. Emilio Coni, prominent hygienist and government official,

had conducted numerous public health surveys that tabulated prison populations, hospital services, legitimate and illegitimate births, and the incidence of prostitution and venereal disease.[3] Homicides, robbery, and other
crimes were cataloged, beginning in 1881, concurrent with the importation
of European sciences of hygiene and social control. The Buenos Aires city
police gathered statistics on murders, thefts, arson, suicide, and street accidents, their numbers collated and published by the municipal government
in an annual statistical journal. In 1888 the city police department began its
own occasional publication of crime data. In the 1890s Juan Vucetich of the
Buenos Aires provincial police, later known for his pioneering work in
fingerprint classification, provided detailed charts and tables of crime. Social pathologists' belief that crime and mental illness were expanding reinforced such documentation as a scientific priority.

Prominent authors, criminologist Antonio Dellepiane among them, attempted to fill some of the gaps in data on crime. In his 1892 book *Las causas
del delito* (The causes of crime), which summarized the Italian and French
schools' ideas on atavism, degeneration, madness, and epilepsy, Dellepiane
had, in a "sociological" section, sought to reconcile biological, social, and
psychological causes. Intending the book to be "exclusively a study of our
own criminality," with two chapters devoted to Argentine crime, he cited
economic and political crisis in Buenos Aires as the cause of attacks on civic
authority, and he claimed violence in rural areas was a consequence of
gaucho activities of banditry, gambling, and drunkenness. His underlying
hypothesis was that immigrants were largely responsible for urban crime in
Argentina.[4] As a result, the capital police began to organize findings in
incident reports by national origin, routinely separating "*argentinos*" from
"*orientales*" (Uruguayans), "*rusos*," "*italianos*," and three or four other national categories. The political orientation of activists became another preoccupation of the statistics gatherers. The police and other observers of
crime, tallying strikes, demonstrations, and other disturbances, classified
labor activism as criminal activity. In the late 1890s they added a new
category, "anarchist violence," to the catchall of "crimes against the social
order." It also was linked to immigrants. Italians or "Russians" (often a code
word for Jews) were said by press and scientists alike to have brought
anarchist doctrines with them from the Old World to the New.

The nation's criminologists sought to measure and control this criminal
world. The underworld and its inhabitants became a subject or "social fact"
to be quantified, known, and controlled. In this effort, the primary nurturing force was the young physician José Ingenieros, who was to become one

of the best-known figures in twentieth-century Argentine history. Born in
Palermo, Italy, in 1877 and raised first in Montevideo, Uruguay, he had
migrated to Buenos Aires as a youth, together with his mother and journal-
ist father, to study at the elite boys' high school, Colegio Nacional. He became the star pupil of two prestigious figures at the medical school of the
University of Buenos Aires, José María Ramos Mejía and Francisco de
Veyga, the one an expert in neurology and the other in hygiene. By 1900
Ingenieros had earned his medical degree, and his doctoral thesis on "The
Simulation of Madness" was honored by the medical faculty's top prize. It
was a Darwinian interpretation of the psychological state of "malingering,"
the feigning of mental illness. On Ingenieros's graduation, his mentor Veyga
appointed him chief of the neurology clinic at the University of Buenos
Aires. In that role he was commissioned by the municipal government
to survey social and hygienic conditions of workers' quarters. Between
1904 and 1911, when he served as director of the Servicio de Observación
de Alienados (Lunatic Observation Service) of the Buenos Aires police,
his reputation spread abroad. The Parisian Societé Medico-Psychologique
named him a corresponding member, and he was selected as Argentine
representative to the Fifth International Congress of Psychology in Rome,
where he served as president of the Pathological Psychology Section. In
1907, supported by the National Penitentiary, he founded and directed its
new Criminology Institute. Ingenieros was a founding member of the Argentine Psychology Society the following year and president of the Argentine Medical Association the year after that. His career, his political activism, and his international reputation exemplified the rising legitimacy and
status of Argentine psychiatry and criminology in the new century.[5]

Ingenieros's mentors, Veyga and Ramos Mejía, were passionate members of Argentina's modernizing Generation of 1880, and Ingenieros himself
was strongly influenced by their unwavering belief in science and by their
social elitism. Yet, early in his career, Ingenieros took up a critical stance to
the left of both Veyga and Ramos Mejía. He had joined the Socialist Party at
age eighteen and was active until 1902 in that moderate reformist party,
which competed for members with a growing radical workers' movement
(especially the anarchists). Long after his departure from left-wing politics,
he retained a basic interest in, and understanding of, the social factors of
crime.

Ingenieros was, at once, a key interpreter of European thought in Argentina and an innovator in his own field but, above all, a builder of institutions

that allowed a growing corps of physicians, psychiatrists, jurists, penologists, and police officials to collaborate in the study of crime, mental illness,
and deviance. United in their scientific ethos, worldview, and approach and
increasingly subsidized with government funds, they, in turn, proved ready
lieutenants for the campaign to professionalize the new science of crime.
The earlier, dispersed nature of studies of crime among hygienists, doctors,
lawyers, and police gradually gave way to further specialization of various
aspects of criminal justice. As in the early 1880s, physicians, such as Ingenieros, dominated the next generation of Argentine criminal anthropology.
As a result, the most prevalent approaches to the study of crime were
"clinical" examinations of individual offenders, with an emphasis on laboratory studies, patient observation, and treatment recommendations. Ingenieros championed the application of Social Darwinist theories to Argentine problems. He aimed to study crime, for instance, within the "struggle
for existence" and published discussions of Herbert Spencer and other Social Darwinians as editor of the nation's major criminology journal, the
Archivos de Psiquiatría, Criminología, y Ciencias Afines (Archives of psychiatry,
criminology, and related sciences).[6] He was an avid consumer of the latest
developments in psychology, the science of human behavior that had grown
out of nineteenth-century "alienism" and "moral medicine."[7] Though not a
follower of Sigmund Freud, Ingenieros was more receptive than his intellectual forebears to psychological explanations of socially disruptive behavior.

One of Ingenieros's signal contributions to the discipline was the 1902
founding of the *Archivos*. In the pages of the *Archivos* this new generation
of criminologists redoubled its predecessors' efforts to interpret European
criminological theory. They defined the main themes and ideas that were to
guide their statistical and descriptive studies of deviance. Ingenieros hoped
his journal would bring students of crime together for the exchange of
specialized ideas. He envisioned it a repository for new research findings,
as a platform for the new Argentine criminology, and as a forum for international scientific debate. Its prestigious pages, for the first time, provided
in Argentina a unified ground for criminological science and expanded the
space devoted to publication and discussion of local crime data. It quickly
became Argentina's leading vehicle for European theory, the chief arena of
internal debate and innovation, and ultimately, a prized exemplar of scientific excellence for the nation. For many years it was the only Latin American journal exclusively dedicated to criminological science.[8]

The *Archivos* was understood as the successor of *Criminalogía Moderna*, the left-wing publication edited in Buenos Aires in the late 1890s by Italian journalist Pietro Gori. The new journal, however, would not suffer the financial difficulties of its predecessor. It was subsidized, housed in the national penitentiary, and produced with prison labor. Whereas Gori saw economic inequality as the root cause of criminal behavior, Ingenieros looked at the inner qualities of the criminal—morphology and psychology, an approach compatible with state goals of confinement and segregation of disorderly individuals. The state rewarded Ingenieros handsomely for assisting in making the process more modern and professional and thus, they suggested, more humane. While Ingenieros advanced rapidly to ever higher government positions, including a stint as assistant to President Roca, Gori left Argentina for Chile and eventually returned to Italy, his legacy of socially conscious criminology largely overshadowed by the new biopsychological model.

A platform for the dispersal and exchange of peer-reviewed research, the *Archivos* provided the essential ligaments of a developing body of research on Argentine crime. Indeed, unlike the short-lived journals and isolated monographs of the 1880s and 1890s, it supported a sustained conversation on crime among Argentine, Latin American, and even European scientists. Ingenieros wrote that "The *Archivos* will try to establish the special ways in which we in the South American continent are reviewing the phenomena of individual and social psychopathology, thereby completing the studies of European researchers."[9] The editors of the *Archivos*, notwithstanding their desire to transcend any particular school of thought, and like their mentors of the Generation of 1880, were drawn to a particular set of theories: the Italian positivist school. In addition to its regular reviews and critical essays on Cesare Lombroso, Enrico Ferri, and others, the *Archivos* devoted almost an entire volume in 1906 to Lombroso on the occasion of his retirement. The appeal of the Italian school was enhanced by massive Italian immigration, which led to more extensive cultural and physical links to Italy, a prolific circulation of Italian books and newspapers, and frequent exchanges between Italian and Argentine intellectuals.[10] Many of the principal figures in early Argentine criminology, including Gori and Ingenieros, were themselves Italian-born. Simultaneously, while not abandoning idealization of northern European culture, Argentine intellectuals and political leaders began to articulate an incipient pride of their own, a kind of "*latinidad*" or "Latinness," which led them to reexamine theories developed by other southern, especially Spanish and Italian, thinkers, hitherto rejected or marginalized as representing inferior thought.

A Synthetic Program of Psychopathology

In the pages of the *Archivos* after 1900, Ingenieros and colleagues proposed a comprehensive new methodology, an "Argentine school" of psychopathology, that sought to make their assessments of abnormals more precise and calibrated to specifically Argentine deviant types. They shared the eclecticism and the concern about crime and mental disturbances of their predecessors but repackaged them as part of an accelerating professionalization of their discipline. In doing so they believed they could transcend the supposed parochialism of European schools of thought with a contribution that transcended, in their view, defining but necessarily narrow approaches. This would, in turn, they speculated, transform their largely derivative science into a fount of special insight worthy of international respect. Ingenieros would claim in 1907, "Until now, criminological studies have passed through their period of evolutionary formation. . . . [I]t is now possible to sketch a synthetic program of criminology."[11]

Argentine criminologists were convinced that their eclectic approach was the best way to deal with the complex Argentine reality, that the polyglot, hybrid nature of their society demanded a flexible approach open to every theory or method, including Darwinian models, Italian anthropometrics, French psychiatric theories, and the theories of some who cited environmental causes such as poverty, child abandonment, and alcoholism as the causes of crime, vagrancy, and begging. According to Ingenieros, in order to defend Argentine society properly, its professionals needed a tailored theory of criminality fit to its own national problems. His approach to the causes and treatment of crime merged biological, social, and psychological factors. He called that approach a "psychopathological school," announcing it as a "new school" of great importance. His emphasis on psychopathology, nonetheless, resembled Ferri's "criminal temperament."[12]

Partially rejecting Lombrosian biological determinism and anthropometry, Ingenieros in his criminology writings of the early twentieth century placed greater emphasis on the social and psychological factors of crime. With a physiological base, he suggested, "one sees that the barriers between psychologists and psychiatrists are disappearing little by little. One essential difference remaining between them is therapeutic practice; but, even here, it is appropriate to point out that psychologists tend to convert themselves into doctors of the spirit."[13] One Italian contributor to the *Archivos* agreed: "Psychiatry occupies an important place between the medical, psychological, and social sciences. . . . On the one hand, it connects [*anastomo-*

sar—a biological term] with general clinical investigations and with pathology; on the other hand, it touches the lofty problems of the spirit and human actions."[14]

While the *Archivos* adhered to the Italian positivist school in inspiration and purpose, its contributors pointed to deficiencies in Italian theory and sought to transcend them. Ingenieros and his colleagues criticized, shaped, and sometimes even rejected their European models. In the introductory essay of the first volume of the journal Ingenieros wrote:

> Undoubtedly in the first studies of the Italian school, the importance of anthropological factors was exaggerated; but, in sum, it was a useful exaggeration, in as much as the great scientific clear-sightedness of Lombroso and his school consisted precisely to demonstrate that the criminal is abnormal. But soon enough, in Italy itself, the criticism was polishing up the primitive concept, and the Italian school definitively consecrated—by means of [Enrico] Ferri—the existence of three classes of factors in the etiology of crime: anthropological, physical, and social.[15]

In an interesting reversal, Ingenieros even boasted of Argentine influence on the great Italian master himself, noting, for example, that Lombroso's *Man of Genius* frequently cited José María Ramos Mejía's *Neurosis de los hombres célebres en la historia argentina*.[16] Ingenieros argued that his own psychological level of analysis, which built on the work of Ramos Mejía, was more highly developed than European theory: "The first phase of criminal anthropology, as with all new sciences, was empirical. The *morphology* of the criminals was studied before their *psychology* . . . . In its later phase, criminal anthropology attributes to criminal *psychology* greater importance than the study of morphological anomalies. Crime is an action; all actions are the result of an active psychological process."[17]

After 1900, Ingenieros's studies of crime explicitly rejected the biological and neurological studies of his mentors. Although his embrace of Freudian psychology—by then of growing influence in Europe—was far from complete, Ingenieros now incorporated theories of mental states. Ingenieros, emphasizing the psychological component, stated, "The *specified* study of offenders . . . should be that of the abnormalities in their psychological functions. . . . If one can speak of schools to designate scientific tendencies, the new school of criminalogy [*sic*] should be called the *psychopathological school*."[18] As a junior faculty member at the University of Buenos Aires medical school, he offered the first systematic survey of the new methods in criminal psychopathology, in "psychic anomalies in criminals," and the "new psychological classification of criminals."[19] His focus on psychology complemented the positivist tendency toward individual explanations of crime and the idea that treatment or reform must be fine-tuned to the particularities of an individual subject. The psychopathological approach expounded by Ingenieros, so Argentine criminologists believed, took the best of all existing theories and represented a new step up in the evolutionary progress of their discipline. To Lombroso's empiricism and clinical method, they added Ferri's social factors, French sociology, and an emphasis of their own on neurological and psychological research.

A Taxonomy of Delinquents and Deviants

Medical and psychiatric approaches that focused on the individual physiognomy and biology of the patient offered a new method of classifying not only criminals but the mentally ill. In this system of classification, at the most fundamental level, scientists struggled to make sense of the wildly disproportionate role of men in criminal activity and the rare but peculiar cases of female lawbreakers. Others concentrated on the sexual perversities and inversions that seemed to them to play a significant role in provoking criminal acts and that, to their grave discomfort, seemed to define the norm within the all-male world of the prison system.

Classification was at once a remnant of the ancient habit of hierarchical ordering and an outgrowth of more recent positivist or empirical approaches. Scientists had arranged plants, animals, and people into groups or classes since at least the time of the ancient Greeks. The ordering of objects and creatures in the natural world was part of the nineteenth-century taxonomic culture within most scientific schools. Charles Darwin's notion of genealogical descent was taken up by, among others, Ernst Haeckel of Germany and Sir Francis Galton of Britain, who applied it to human groups. In the field of medicine, disease classification, or nosology, had been based largely on symptoms until the introduction of germ theory in the 1870s, but not all diseases (especially mental ones) could be explained under that rubric. Criminologists believed that they had to specify a taxonomy of criminal types before they could draft legal reforms to prevent or treat socially disruptive acts. Only with a firm sense of the different types of criminals—and their different degrees of "dangerousness"—could scientists begin to construct correctives. They hoped to present new tools even to anticipate crime and to prevent criminal behavior by scientifically identifying those destined to act. By defining criminal types, they could shape the laws in advance of crime.

Argentines, like their north Atlantic counterparts, applied classification to large social groups, not individuals alone. Ingenieros argued that one of positivist criminology's main contributions would be the specification of subclasses of criminals, "mental deficients," and other "abnormal" types. He wrote, "In this new classification we do not confuse types or categories . . . . [W]e isolate various heterogeneous types which have until today been confused in one group." He outlined the rationale for classification: "Criminal psychopathology . . . clinically demonstrates the existence of various criminal types in which emotional, intellectual, and volitive anomalies predominate (pure types). This differentiation serves as a clinical basis for the classification of criminals." This new approach "permits an appreciation of the reformability and dangerousness of criminals; each group . . . corresponds to varying degrees of anomalies in the antisocial character. In this sense, [the classification] can be better adapted to the new principles of Criminal Law and modern penitentiary practices."[20]

In 1905 Ingenieros formally introduced his new psychopathological approach to criminal classification in a paper presented at the Fifth International Psychology Congress in Rome and published in the *Archivos* the following year. He began on an optimistic note, declaring that "the science of crime has left its empirical phase and now begins to define some general principles." Acknowledging Ferri's influence in the construction of his classifications, Ingenieros described his own scheme. "In this classification, one does not in any case confuse two categories or groups that are separated in the classifications of other authors. . . . With respect to Ferri's classification, highly superior to all the other empirical classifications expounded until now, we separate the born impulsive from the amoral, the congenitally and permanently insane from the accidental psychopaths, the impulsive passions from the obsessive ones, the occasional abnormals from the occasional impulsives, etc."[21] Ingenieros had constructed a flow chart in which "mesological" or social causes could be separated from "anthropological" ones. In the latter, a more extensive section, Ingenieros separated somatic from psychological factors, assigning each general category its own types of "abnormals." Five years later, Ingenieros would amplify the anthropological causes of criminal psychopathology in an amended chart, distinguishing more precisely between types.

Argentine scientists focused on crime as an urban phenomenon. Rural crimes were acknowledged but considered less significant and less threatening to national development.[22] In part, this was a circular argument: the immigrant "problem," for instance, was largely considered an urban one, and the most disturbing crimes were considered a result of this influx of undesirables. The urban bias was expressed in a number of ways. The overwhelming majority of data collected pertained to city crimes. When scientists did gather statistics of rural crimes, their goal was to compare them to urban ones and not to address rural phenomena, an empirical neglect reflected in state practices as well. In addition, most provincial areas lacked the theoretical and bureaucratic infrastructure that was then reconstructing large parts of Buenos Aires's own criminal justice system.

But, despite the emphasis on identifying urban immigrants, class status was far more significant than nationality in determining criminality. Crime rates for all major ethnic groups remained roughly constant with the representation of each in the population, Spanish immigrants having a slightly higher rate than native-born Argentines, Italian-born immigrants a lower rate. The few scientists who questioned the blame attributed to immigrants were overwhelmed by prolific scientific and popular stereotypes of immigrants. In 1912 Miguel Lancelotti, a well-known criminal lawyer and statistician, recalculated crime rates by age as showing bias against foreigners, but criminologists discounted his critique.[23] The statistics revealed, too, that most types of crime were, in fact, decreasing per capita at this time. The increases that captured the attention of contemporaries were crimes against people and property, such as assault, theft, and vandalism, visible and alarming, which surged during periods of economic crisis.

"A Useful Exaggeration": Classification and Race

Just as modern medicine was discovering congenital, even genetic, proclivities for illness, so too the editors of the *Archivos* seemed fully to accept biological theories of crime, as well as the concept of the born criminal, put forward in the 1870s by Lombroso. A 1906 article by J. A. González Lanuza, a Cuban law professor, stated, "From the combined teachings of the founder [Lombroso] come, in my view, these salient and essential features . . . the doctrine of the *criminal type*, an affirmation of the existence of the *born criminal*, the determination of biological factors of criminality, and the comparison of congenital criminality with so-called 'moral insanity.' "[24] As the concept evolved it would take on racial connotations.

Ingenieros's founding statement in the *Archivos* in 1902 claimed that "born criminals" had a higher proportion of "psycho-physiological" rather than of social factors, but that the reverse was true for "occasional" criminals: "Thanks to the laborious investigations of . . . Ferri and many others, visible and measurable morphological anomalies, deformities, and divergences from the average type, constitute a vast scientific arsenal, from which we can establish criminal biology." However, Ingenieros warned that it was difficult to identify "criminal types" by physiological stigma alone, since they exhibited the "morphological anomalies common to all degenerates." He presented these categories in yet another chart, showing his view of the proportions of biological and social causes of criminality and the varying levels of social "dangerousness" of each type. The most dangerous type, the "born criminals," represented the smallest number in his list. At the other end of the spectrum, "occasional criminals," the largest number, were described as products of their social environment. It was this group, Ingenieros believed, that would be most responsive to reform efforts.[25]

Anthropometrics took off, not just to help stop criminals but to classify and type the general population. Police scientist Juan Vucetich in 1904 evoked evolutionary theory in his description of patterns of fingerprints, seeing evidence of a link between humans and primates. While monkeys, too, were found to have individual fingerprints, Vucetich concluded that certain types of patterns were found more frequently among the lower species. His conclusion provided a link to the more anthropometric interpretations of his science and fed the popular theories of biological determinism at the time. Criminals, epileptics, and other "degenerates," Vucetich claimed, tended to have loops and whorls similar to those of the primates. "The variety of patterns observed by Alix in monkeys was discovered five times by M. Feré in his epileptic patients. This conformity, common in anthropoids and very rare in man, would form a bond uniting the species. And the forms closest to those present in monkeys, according to some authors, would be those of degenerates of all types. . . . Degenerates display with greater frequency the shapes called *primaries*." Vucetich called for further study of the correlation of fingerprint patterns and criminal types. "The frequency of a determined form," he wrote, "serves as a point of departure for future investigations of anthropologists and medicolegists." What is curious about this stance is that, a few pages later, Vucetich denied the influence of heredity, race, climate, and "level of civilization" in determining the pattern or shape of individual fingerprints. Nevertheless, in evoking the links between fingerprint patterns and degeneracy, he demonstrated the continuity in thought between a medicalized criminology and the more "objective" technology used in criminal identification.[26]

Discussions of foreign criminal crowds began to be interwoven with assumptions of racial inferiority and stigma. In 1888 Luis Drago, criminologist and judge, equated criminals to members of primitive societies, labeling both enemies of civilization. Criminals were similar to "savages," as documented by comparing "physiognomic and somatic similarities" and other signs of atavism.[27] As evidence, Drago included an appendix listing comparative measurements of brain weight: "In Europeans, the average weight of a man's brain . . . is 1,403 grams, [and] a woman's is 1,247 grams. In idiots one finds brains of 765 grams, 730 grams, 560 grams. . . . The average normal brain weighs 1,362 grams." He cited the composite facial photographs of criminals compiled by Herbert Spencer and Sir Francis Galton in England as further evidence of observable and measurable signs of degeneration.[28] Eusebio Gómez, director of the National Penitentiary, too, pointed in his discussion of crime to an 1881 book on the natural history of the northern Chaco region of Argentina, which described "numerous measurements of Indian skulls, taken from the *toba*, *chunupí*, *chiriguana*, and *payaguá* tribes." The book included an extensive list of cranial measurements and line drawings of "typical" native features.[29]

The medicalization of criminal behavior was reflected in the central role of legal and forensic medicine in the *Archivos*. Studies of legal medicine, conceived as a bridge between medicine, psychiatry, and the law, appeared in the form of autopsies of famous or otherwise notable criminals, like assassins or grisly murderers. Medicolegal case studies reflected anthropometric methods, with descriptions of subjects' physical features and of any diseases they might have suffered, such as epilepsy, tuberculosis, or syphilis. The autopsy report of the murderer Mateo Morral, published in the *Archivos* in 1907, read, "One notes in the right ear a sign of degeneration."[30]

The Argentine criminologists were most enamored of Lombroso's emphasis on clinical method and, in particular, his practice of observing and experimenting on pathological subjects. Francisco de Veyga wrote in 1906 that Lombroso's "very important accomplishment" was the "observation and presentation of the phenomena which have motivated his theories."[31] He credited Lombroso with initiating clinical methodology in criminology and proclaimed Argentina's great debt to his positivist movement, whose goal, Veyga wrote, was to understand the "morbid types . . . the monstrosities, inconsequentials."[32] To J. A. González Lanuza, a professor at the University of Havana writing in the *Archivos*, Lombroso's anthropometrics was "the only scientific method."[33] Numerous other authors referred to morphological traits and physical stigma in theoretical and medicolegal reports in the *Archivos*. In 1906 Veyga, writing in homage to "Lombroso's *oeuvre*," noted that "crime is an atavistic phenomenon; Lombroso found it in animals, in primitive man, in the savage. . . . Crime and the criminal appear, then, in the penetrating view of Lombroso, as living reflections of the species' past, springing from hereditary degeneration."[34]

Ingenieros's first announcement of the appearance of the *Archivos* had lamented the remnants of "classical" approaches to crime in Argentine law and promised the new outlook on crime as a stark improvement over the older views. As Ingenieros explained at the time, "The classical school of penal law—whose spirit abounds in contemporary legislation—considers crime as a simple juridical act. It does not attribute importance to the organic and environmental conditions that contribute to its determination. Crime appears as an abstract entity, independent of all determinism, susceptible to punishment as an expression of the evil intents of the offender."[35] An accurate and modern criminology would focus on individual criminals and their traits, thereby imbuing the law with a "new life, more intense and fruitful, more true." Ingenieros cited the "discovery" in medicine that there "are not illnesses but patients." Just as a "true doctor" does not have the same treatment for each illness, so a "criminologist knows that in each case he must do a special study and refrain from applying an *a priori* formula." But ideas about who was deemed necessary to examine was the engine that drove these new clinical studies, and not the reverse. Ingenieros described the *Archivos*'s central goal as the "scientific study of abnormal men [*hombres*], especially criminal man and the insane."[36] He began to observe Argentine prisoners, mental patients, and other individuals to construct theories of criminality that were supposedly tailored to Argentine special realities and needs.

Foremost, so the new professions maintained, was society's ability to defend itself, which rested on measurement of degrees of "dangerousness." Therefore, places needed to be found to put the methods of anthropometry and psychopathology into practice, for psychiatrists and criminologists to identify and diagnose criminal types in order to recommend appropriate punishments or treatments for each. The state provided such places, investing unprecedented funds in these new techniques and for structures of social containment of disorderly residents: criminals, mentally ill, vagrants, prostitutes, anarchists, and urban "lowlife." It reinforced the ability of the police to classify and contain such persons. It supported a research program to propose a comprehensive follow-up plan for police, prisons, and the legal justice system.

The earliest of these endeavors was the Lunatic Observation Service (Servicio de Observación de Alienados) at the capital police station, founded in 1899 in Buenos Aires. There, scientists prepared biological, physiological, and psychological profiles for each subject examined. The observation room was designated as a stopping-off point for the men and women picked up by police officers because they acted suspiciously or seemed mentally ill. Family members, too, could bring in relatives who were behaving strangely. This clinical station took on a new professionalism with Ingenieros at the helm. He brought his expertise to bear in diagnosing mental disorders and distinguishing "simulators" from the truly ill. He saw in his new job ample opportunity to develop further his skills of diagnosis, analysis, and classification. Ramos Mejía wrote that clinical cases were the "required basis . . . for those generalizations which allow us to induce from the study of individual cases the nosological physiognomy of illnesses."[37] Ingenieros indicated that "clinical criminology," as practiced at sites like the observation room, would provide fundamental scientific insight: "The clinic offers us examples of profoundly degenerate individuals, in whom coexist impulsiveness, the absence of moral sensibility, and disrupted mental functions." Two years later, Ingenieros added that his continued clinical examinations represented "a brilliant conquest by the medical tendencies that have renewed the very fundaments of penal law, crowning the work begun by alienists and criminologists."[38] The entry of criminologists in police stations and, later, in the prisons signaled their acceptance as professionals in state institutional settings, and the partnership was further cemented by shared assumptions about the importance of "social defense." Criminologists positioned themselves, and convinced government officials to accept them, as uniquely qualified to provide the essential clues for identifying, treating, and preventing criminal behavior.

A central observation that emerged from early clinical and theoretical work alike was the presence of a racial problem in Argentina that was linked both to the nation's problematic Indian past, as well as to the present challenges posed by immigration. Following European trends, Argentine criminologists considered criminals in racialized terms, as "degenerate" or "deformed," a concept that implied a set of racial categories of particular meaning and consequence to Latin Americans. Most European scientists, who themselves promoted a doctrine of racial purity, found miscegenation—of which Latin America, including Argentina, was a prime example—extremely troubling. Similarly, most Argentine intellectuals were deeply preoccupied with the links between progress, civilization, and racial mixing. But there was a range of often contradictory thought about the meaning of race for Latin American nations. While most Argentine scientists held the Anglo-Saxon and Nordic "races" in highest esteem, they were unwilling to succumb completely to pessimism about their own "Latin" and Hispanic nation, ascribing to—often in the same breath—ideas about the correctability or mutability of national character traits. Some, like Ingenieros, believed that no significant distinctions (or hierarchy) existed among Europeans and reserved their harshest judgments for Indians and Africans.[39]

Racial assumptions permeated criminological thought of the period. Scientists used the term "race" to refer not only to phenotypic categories of people but also to supposedly distinct ethnic or national groups (for example, the "Latin" race). Victor Mercante, a prominent educator, wrote in 1902 that "D", an Italian immigrant offender, "owed his character nearly exclusively to the anthropological factor, that is to say, to the spirit of his race."[40] But despite the centrality of the idea of race, the manner in which they used it was vague and ill-defined. It could connote different ethnic categories or biological differences. It could refer to nationality, but it was often used to link biological and cultural factors of different racial groups. In Argentina itself, race usually meant ethnic, or even regional, types, but journals such as the *Archivos* occasionally published studies about racial groups in countries with higher numbers of people of African descent, such as the United States, Brazil, and Cuba.

Lucas Ayarragaray, a physician and national legislator who contributed to the *Archivos*, suggested in 1912 that the central problem facing Latin American nations was racial, "the anthropological dilemma" of their "ethnic constitutions." While Ayarragaray accepted immigration as necessary for economic reasons, he felt deeply threatened by the "lower" races and the prospect of miscegenation, and yet at the same time he promoted the "whitening" of darker races. In Ayarragaray's view, Indians and Africans would not begin to improve themselves until their "dark" blood had been diluted by two-thirds by white blood. He considered any form of mestizo a degenerate. Even those of one-quarter African or Indian ancestry, he said, suffered from numerous physiognomic stigmas, such as abnormal foreheads, limbs of uneven length, and weak constitutions.[41]

Race was also considered to encompass Jews, a religious community and a significant minority in immigrant Buenos Aires. Psychiatrists and criminologists of the early twentieth century contributed to anti-Semitic discourse by publishing articles citing Jews as a distinct racial group. In 1910 an article proposed that Jews displayed a supposed vulnerability to certain mental diseases. The *Archivos* reviewed a study from the French journal *Archives d'Anthropologie Criminelle* that supported the common stereotype that "among the Jews, to tell the truth, one observes a higher frequency of intellectual disturbances, which is a consequence of their generally greater intellectual activity." The author cited European studies, including those of Lombroso and French neurologist Jean-Marie Charcot, as concluding that Jews were more vulnerable to neurasthenia, hysteria, neurosis, and "psychosis in general."[42]

Ingenieros and the *Archivos* thus set up a lasting legacy in Argentine human science, establishing a tendency to classify by human "types." When, in 1913, Ingenieros resigned his posts at the Criminology Institute and the *Archivos*, he left behind a strong and unified criminology profession. Two years earlier, he had been passed over for the position of chair of the Department of Legal Medicine at the University of Buenos Aires, the victim of a political deal made between President Roque Sáenz Peña and the Catholic Church.[43] Angry and disappointed, he left Argentina for a three-year tour of Europe, where he studied and lectured in Paris, Geneva, Lausanne, and Heidelberg, met with European colleagues, and continued to publish on criminological and philosophical topics. After his return to Buenos Aires, he virtually ceased all criminological work (though he did revise his textbook *Criminology* a number of times) and focused his energies instead on political writings and questions of national character, such as in his 1913 book *El hombre mediocre* (Mediocre man) and his 1917 *Hacia una moral sin dogmas* (Toward a morality without dogma). By then, the *Archivos* was running smoothly on its own, and developments to bolster state powers of surveillance and confinement were well underway in policing, prisons, and the law.

Ingenieros's successors at the *Archivos* conducted the changing of the guard in a mood of optimism. They announced a new journal, the *Revista de Criminología, Psiquiatría, y Medicina Legal* (Review of criminology, psychiatry, and legal medicine), whose new title attested to the strength of criminology. In the last issue of the *Archivos*, the editors, now under the leadership of psychiatrist Helvio Fernández, announced that the *Revista* would "pursue the same program of study, examination and criticism" as had the *Archivos*. They hoped Ingenieros would continue to act as the "core of an intellectual movement, the propulsion of scientific progress," but, if he chose not to, his legacy of the past twelve years remained for them to continue. "The *Archivos*," they wrote, "is closing its circle with a rallying cry for the future."[44]

The Criminology Institute would continue to carry out "investigations of the abnormal individual, especially the criminal and the insane, in order to establish a basis for the shaping of preventative or sanitary systems."[45] The opening issue of the new journal spelled out those practical goals in more detail: to "bring together all the diverse studies of the abnormal individual," including an investigation of the causes of abnormal psychology, in order to place such individuals in a "determined clinical classification." Finally, the new editors' goal was to "determine treatment, within the therapeutic principles of modern Psychiatry, in special cases of the criminal and the insane, to facilitate the subject's rehabilitation or improvement, or to prevent his possible dangerousness."[46]

Argentine criminologists felt they had achieved a better understanding of the causes of crime and had arrived at some basic diagnoses of the nation's most pressing problems. As it matured, Argentine psychopathology ascended to a critical national role in the introduction and promulgation of legal and social reforms. The editors of the *Revista* continued Ingenieros's emphasis on a relationship between criminology and the law, presenting their clinical work as the basis for legal and penal reforms, their goal being "the solution of the conflict between the criminal subject and society."[47] By 1916, in the sixth edition of his book *Criminology*, Ingenieros suggested that criminology had matured sufficiently to fulfill its goal of reforming Argentine law. He wrote, "Today it is possible to fix a systematic plan to this science, transforming primitive criminal anthropology into a criminal psychopathology, and specifying the social cost of criminal conduct. On these premises we have attempted a new classification of offenders, illustrating the diverse groups with clinical observations and pointing to the practical agreement with the Penal Law in formation and with the new penitentiary tendencies."[48]

By 1914, Argentine social pathologists were nearly convinced that they had provided a unique contribution to the science of crime. Yet though their statistical abilities gained a modicum of attention from scientists abroad, their psychopathological school was taken for what it was: a regurgitation of European theory. Prestige for "national science" would not come from the elaboration of new theories. The effects of state-sponsored studies of crime would be principally felt internally. But, first, the social pathologists had to provide a detailed explanation of the range of symptoms manifested, and from this effort emerged a pantheon of criminal types, disruptions, and other social ills, of "degenerates" that plagued the city and the nation.

FOUR

Defects of Organic Constitution

Degeneration of the Nation's "Germ Plasm"

> Atavistic degeneration breeds forms that are not human but concern inferior
> animals. . . . But there is also a human atavism that reproduces the inherited
> structures of an especially morbid type. Primitive degeneration is the result of
> the struggle for existence during fetal life or in the first years in the world. It
> demonstrates a diminished individual resistance of the exterior conditions of
> life. Acquired degeneration is born in the course of life without inherited or
> embryonic cause. These causes of delinquency can be reduced to the
> retardation of development.
> —FRANCISCO RAMOS MEJÍA

In the late 1880s Argentine scientists and social scientists discovered an explosion in "degenerative" conditions among the Buenos Aires population. New plagues of "atavism" and "biological anomalies" were reported. When the statistical crime tables—the nation's first—were published, they were accompanied by lurid portraits of "degenerate" types, male and female, from murderers to prostitutes and transvestites. In 1888 *Hombres de presa* (Men of prey), written by criminal judge Luis María Drago about the many "debased and degraded beings" he had encountered in his court, was hailed as the first scientific study of Argentine criminal types.[1] Drago, later Argentina's minister of foreign relations, would build his career on cataloging Argentina's proliferating degenerate types. Ten years after *Hombres de presa* appeared, Eusebio Gómez, the noted criminologist and director of the National Penitentiary, updated Drago's work with a book of his own, *Mala vida en Buenos Aires* (Lowlife in Buenos Aires). It expanded the diagnosis of degeneracy to embrace prostitutes, pimps, homosexuals, beggars, vagrants, and charlatans. Said José Ingenieros, the top criminologist of the time, "the entire spectrum of degeneration, in its corrosive and antisocial forms, parade through the pages of this book."[2]

Throughout Europe and the Americas at the time, scientists and politicians used the term "degeneration" to evoke decadence and decay.[3] In an intersecting of biological and environmental causations, modernization began to be seen as spawning degeneration, a decline in or a deviation from some real or ideal human form. Its first systematic application to medicine and psychiatry appeared in the book *Treaty on Degeneration*, by the French psychiatrist Bénédict Augustin Morel, who had earlier identified syndromes of "cretinism" and "dementia praecox" (later renamed schizophrenia). In 1863 Morel had set up a photographic studio in Paris to make composite facial portraits of asylum inmates. By the late years of the century, the term "degeneration" referred to prostitutes, criminals, the insane, and the poor and to a wide range of diseased conditions from epilepsy and neuralgia to retardation, deafness, blindness, myopia, and stuttering, to impotence and sterility, insanity, sexual perversion, alcoholism, and eccentricity. Physicians in Argentina, as elsewhere, used the term not for individual diagnosis alone but to refer to the health of the national body. They were convinced that society itself was sick, most of its symptoms relatively new outcomes of the country's newfound prosperity, bustle, and overcrowding—in short, the result of progress and modernity and yet, at the same time, inconsistently, genetic in cause.

Degeneration was considered both biologically and environmentally based. As Argentine scientists Francisco de Veyga and Juan C. Córdoba pointed out, the "professional criminal," though "endowed with the ordinary aptitudes" and "under the veil of apparent and almost brilliant normality," displayed a "pronounced disequilibrium in all the faculties [with] the absence of certain sentiments (especially moral ones)" and with the "sensory and motor aberrations, strange ideas, and many other phenomena of this type."[4] Physicians believed that environmental "poisons," such as use of alcohol or even poverty, could become, in effect, genetic, to be passed on to future generations through the weakening of the "germ plasm," causing criminal behavior, mental illness, and numerous other anomalies. In the discussion of social problems in this period, degeneration was seen as primarily biological in origin. Even distinction in heart rhythms and pulse rates could be identified in criminals and the mentally ill, scientists affirmed.[5]

Yet, Argentine physicians, perhaps more than those in north Atlantic countries, emphasized some mutability of degenerative symptoms. One of the first dissertations on Argentina's "social pathologies," written in 1891 by medical student Benjamin Solari, reflected the impact of Social Darwinism. Solari, a psychiatrist, would soon direct the influential Buenos Aires medical journal *Semana Médica*. Vigorous medical diagnosis might thwart hereditary "degeneration," Solari wrote. The "decadent era" had begun, observed in "the disequilibrium of the various functions, the immoderate exercise of some in relation to the others, fatigue, excess, dulling of the faculties," and an "accumulation of circumstances [that had come] to form a decadent type." The decadent type "transmitted its characters and modalities through heredity," forming "a link in a chain that should end."[6]

Unnatural Sex: Female Hysteria and Other Psychoses

The *Archivos de Psiquiatría, Criminología, y Ciencias Afines* (Archives of psychiatry, criminology, and related sciences) recorded a wide range of male and female sexual anomalies in Argentina, from sterility and impotence to homosexuality and sexual psychosis. But it gave special attention to female degeneration, which it attributed primarily to sexual and reproductive organs, and to the diagnosis of hysteria, which it took to be a manifestation of Argentine "modernity." The cause was sexual. "When in the presence of a hysterical woman," wrote one physician, he "should never neglect to examine her genital apparatus, because we see with increasing frequency that there exists the productive source of hysteria."[7] An unsigned article in the *Archivos* on "Sexual Functions, Madness, and Female Crime" similarly considered sexual organs the cause of female deviant behavior: "The correlation between madness and disorders of the genital apparatus is evident. . . . We should never forget the influence of menstruation on the evolution of periodic madness (*mania*)," whose symptoms included "sadness, moral depression, convulsive attacks, hallucinations, etc."[8]

José Ingenieros, the nation's leading expert on hysteria, on examining a seventeen-year old female patient—"Argentine, unmarried, Catholic, and occupied with domestic tasks"—at the San Roque mental asylum, found her oversexed and mentally degenerate. From a poor family with a "nervous" and sometimes alcoholic father, she displayed a sexual abnormality that Ingenieros attributed to masturbation. She relieved this sexual abnormality through "unnatural" means, the symptom being abnormal laughter. During and after self-gratification an "irresistible desire to laugh hysterically seized her." Ingenieros wrote, "Since puberty her sexual instinct has manifested itself intensely: unable to resist its pull, she systematically satisfied her sensuality through daily titillations of the clitoris. This produced complete voluptuousness. This habit . . . had a close relationship with her first paroxysm of hysterical laughter." Her first hysterical fit had occurred eight to ten months after the onset of menstruation.[9]

Hysteria in women, a fashionable subject of scientific and medical study in Europe, especially in France, since the 1860s, had become the most common psychiatric diagnosis for women by the late nineteenth century, but in Argentina the first clinical studies did not appear until the early twentieth century, largely the work of a small group of physicians, neurologists, and emerging criminologists. Ingenieros wrote the most extensive study of hysteria among Argentine patients, *Histeria y sugestión* (Hysteria and hypnosis), published and serialized in the *Semana Médica* in 1904, and as editor of the *Archivos* he published dozens of articles by others about the same condition.[10] Hysteria became a medical metaphor for willfulness, volatility, emotion, and a lack of self-control. It was often cited as prevalent, even epidemic, among Buenos Aires women of every class and as a common side effect of modern urban stresses.[11] One author wrote that "half of all women are hysterics, and in Buenos Aires, our professor of nervous diseases [Dr. José María Ramos Mejía], believes the number is even more." The author concluded that hysteria was "one of the varieties of women's character."[12] Women were more likely to suffer from hysteria, said another observer: "Hysteria and its paralysis . . . tend to manifest themselves generally in weak, exhausted individuals or those consumed by excess or vices."[13] One physician wrote that "in a word, woman is more fragile and impressionable."[14] Hysterical women were believed to be unusually suggestible, with a corresponding lack of control over thought and action. But cure was possible, said some physicians, as did Ingenieros, who suggested that the novel techniques of hypnosis could remedy the biological weakness. An article on three hysterical women by Nicolás Vaschide, director of the Experimental Psychology Laboratory in Villejuif, France, that appeared in the 1906 *Archivos* described the first of the three women as an "impulsive degenerate" who exhibited both hysterical fits and "genital impulses," shouting "Men! Men! Men!" for hours and lunging for the testicles of any man who entered the room. The second displayed an "exaggerated sexuality," and the third "had very intense exotic desires. When a man enters the room her eyes shine, she blushes, and with her tongue or other acts she offers her favors."[15]

Overt female sexuality, "nymphomania" or exaggerated sexuality, in women was believed to upset the delicate social balance. Ingenieros discussed a case of "premature sexuality" in a girl who began masturbating openly at age six or seven. In her adolescence her parents called for the police and sought legal intervention to prevent shameful behavior with boys. Her acts "reveal[ed] abnormalities." Another woman, "a hysteric," had committed adultery because of her belief that her husband could not produce children, and she experienced a hysterical pregnancy with a diagnosis of acute depression. Such cases "of rape, adultery, seduction of minors, etc.," Ingenieros pointed out, raised "social-moral" questions as much as clinical ones.[16] In 1912 the psychiatrist Bernardo Etchepare traced a female patient's precocious sexual activities, including masturbation, bestiality, and reading pornography, not to inborn degeneracy but to abuse by her older relatives. She had taken on a "dominating masculine character, [she] preferred boys' amusements: soccer, running, climbing trees," and with the onset of menstruation she had become depressed and angry. Her teenage years were characterized by sexual activity, periods of depression, and suicide attempts. As hysteria and drug abuse took their eventual toll, her "male character" became more noticeable: she began to ride a horse and a bicycle and shoot guns. The examining doctor concluded that she was doomed to unhappiness, incurably mentally ill, probably a "psychosexual hermaphrodite."[17]

All the functions of a woman's body, potentially pathological, could cause aberrant behavior in women. Physicians in the new state-sponsored observation posts studied menstruation, pregnancy, lactation, and menopause for their effects on women's mental condition. One of the earliest theses on pathological female reproductive functions, "Puerperal Madness" by Fernando Raffo, published in 1888, identified four distinct periods when such madness could appear—during pregnancy, labor, postpartum, and lactation. The majority of cases, 58 percent, were said to occur during the postpartum stage, fewer during lactation and pregnancy; and multiple pregnancies could have a debilitating effect. But despite the link of madness to circumstance, Raffo, like most physicians at the time, considered heredity a cause: "It is certain and admitted by all medico-alienists that heredity is one of the predisposing causes of cerebral disturbances during the puerperal state."[18] A 1917 thesis by Carlos Fonso Gandolfo, entitled "Puerperal Psychosis," defined a woman's puerperal stage broadly, beginning with conception and ending when a woman was fertile once again, and described puerperal psychosis as corresponding with pregnancy, birth, the postnatal period, and lactation. Written with an impressive imprimatur, under the direction of Domingo Cabred, vice dean of the medical school and president of the Argentine Academy of Medicine, Fonso Gandolfo's thesis argued that pregnancy was inherently pathological and "always produces intoxication in the woman."[19]

As in other analyses of personal or social pathology in Argentina, however, some linked degeneracy and madness to the environment. A 1903 study, "Madness during Pregnancy" by Eliseo Cantón and Ingenieros, which stated that "the clinical relationship between pregnancy and madness, known since Hippocrates, has been the object of interesting studies in modern psychiatry," saw women's sexual psychosis, much like crime in men, as an inevitable effect of modernity. "At the turn of this century," Cantón and Ingenieros wrote, "when life is so hard and complicated, it is not strange that a pregnancy provokes in women disturbances of the nervous system . . . . [T]he effects of this puerperal pathology will only increase with the demands of civilization. In this situation, the mother's weakness is exaggerated; previously transitory disorders make themselves definitive and disturb for the rest of her life."[20] But listing as related syndromes "lactation psychosis" and "menstrual psychosis," they traced the tendency of the latter in women with a personal or parental history of hysteria and neurasthenia and considered it an inherited mental condition like depression, mania, and eroticism.

Such pathology among women was not only individual and not found only in the home. Adolfo Batíz, a Buenos Aires police officer with a literary bent, reflected on the range of degeneracy that he had observed among women and men on the streets of Buenos Aires: "The degeneration of homosexuality, like the exercise of female prostitution and degeneration in general, has taken truly exceptional proportions [that are] only comparable with the times of imperial Roman decadence." Prostitution, at its peak in Buenos Aires, could destroy the family, weaken the economy, and destabilize society, Batíz warned.[21] Eusebio Gómez, in his *Mala vida en Buenos Aires*, found the roots of prostitution in "external causes" such as poverty or exposure to bad morals, but in some women prostitution had a genetic cause, "a defect of the organic constitution." He thus illustrated his point: "We knew a family in which, from the grandmother down, all the women were prostitutes." Even the women accepted the family tendency toward prostitution. In an interview one of the women repeated a ribald Spanish saying that ran in part: "Mother a whore, daughter a whore." Prostitutes, said Gómez, were "sowing here and there the malignant germs that infect the body and soul of men." He pointed out that in the dangerous underworld, prostitutes socialized in the same bars and cafés as anarchists, who counted them among the victims of the capitalist state.[22]

A 1912 study of female criminality by H. Leale, "Criminality of the Sexes," published in France and reprinted in Argentina in the *Archivos*, found women neither more nor less prone to criminality than men but liable to commit different types of crime.[23] The author rejected the argument, then growing in popularity, of a superior female moral sense. The Italian positivist school (in contrast to the classical school), he said, maintained that women were more criminal than men and that there were indeed more women criminals, that is, if, as Lombroso pointed out, prostitutes were considered a type of criminal. But Leale referred to the work of Pauline Tarnowski, one of the first women criminologists in Europe and an expert on female criminality in turn-of-the-century France, who cited anthropometry and craniometry's measured typing of skulls as evidence that women were more moral than men but physically weaker and hence less prone to crime: "Woman lives her life in the home. . . . The inferiority of her physical strength obliges her to avoid fights, wounds, and disputes, which are more frequent in men." Leale saw the cultural context as contributing to criminality in the Baltic states of Russia, where women who worked in the fields alongside men, he said, were more likely to commit crimes than women who worked at home.[24] (Victor Mercante, an educator who conducted research on deviant behavior in schoolchildren, disputed the common opinion that women were less intelligent than men: "she is not inferior, she is *different*." Yet he cited phrenological studies that found men's brains to be heavier than women's.)[25] Antonio Dellepiane suggested that in Argentina, where more women were working outside the home, crime was rising precisely because immigrant working women were leaving children home alone, leaving their abandoned daughters to be raped. (Native-born Argentines of Spanish descent, on the other hand, he claimed, seldom worked outside the home.)[26] Women were also, in most cases, held solely liable for crimes of abortion and infanticide.

Criminologists' obsession with prostitution and infanticide ignored the evidence that women who were most often imprisoned in Argentina were convicted of theft. But still criminologists linked theft to inherent female traits; domestic servants who stole from their employers were mentally ill or morally flawed.[27] Domestic servants, predominantly female, while "indispensable" to modern society, were "dangerous to property." Two physicians for the municipal government called kleptomania a female symptom related to mental disorders. In studies in the *Archivos* they linked petty theft to other compulsive behavior in women, like pyromania, dipsomania (alcoholism), smoking, and drug addiction, all disorders whose cause was an inability to control behavior. The urge to steal was a psychological ailment, a "mental pathology . . . a psychic disturbance and not a shameful vice." They cited as a particular cause of theft the delirium of pregnant women who sought out unusual foods to satisfy their appetites. Hence, pregnancy, too, could be a determining cause. An article by medicolegal expert Jorge Coll in the *Revista de Criminología, Psiquiatría, y Medicina Legal* (Review of criminology, psychiatry, and legal medicine) also related pregnancy to kleptomania: "Among the diverse manifestations originating in the alterations of character during pregnancy . . . one of the most frequent is this form of kleptomania."[28]

Most important in the eyes of Argentina's behavioral scientists, for prostitutes, female offenders, and the mentally ill the real crime was the rejection of maternity and bourgeois ideals of domesticity. Doctors and researchers, led by the progressives of their time who espoused this view, may have hoped to help their women patients by enhancing their acceptance of their gender roles and, at the same time, to serve society's needs, but in the process, they reduced their female patients to their sexual and reproductive organs. This research began with an assumption of psychological sexual difference and led to studies that confirmed that difference and diagnosed it as pathology.

At the same time, medical science turned to examine what its practitioners viewed as sexual anomalies in both sexes. A 1904 article in the *Archivos* had labeled hermaphroditism and also homosexuality as forms of "psychic degeneration" and homosexuals, lesbians, and transsexuals as a virtual "third sex" that destabilized accepted gender roles.[29] José Ingenieros linked homosexual behavior in women to hysteria and other mental diseases. Certain types of "independent" women, such as artists and intellectuals, were likely inverts.[30] Well-known educator Victor Mercante spoke of the supposedly widespread lesbian sexual activities in girls' boarding schools as an "epidemic," and he warned that such activities could lead girls "on the road to hysteria."[31] However, these "conditions," frequently termed "sexual inversion," were, unlike hysteria, believed to be more common among men.[32] In a long 1879 article on "passive pederasty," Benigno Lugones, member of the Argentine Medical Circle, depicted homosexuals as assuming stereotypical female roles: "Pederasts iron, sew, weave, arrange their rooms with an entirely feminine coquettishness."[33] Another article in the *Archivos* wrote of a male "invert" who was experiencing secretions from the breast. He was reported to say, "I am weak like all women. I am not able to do anything that other men do . . . . I can't work like them. I like to sew, clean, iron, and cook."[34] Francisco de Veyga, in a study of inversion, wrote of another male subject: "His mental state is completely feminine: cowardice, fickleness, submission, and sensitivity."[35]

Physician-researchers infiltrated the city's homosexual subculture to study its way of life. De Veyga observed, examined, interviewed, photographed, and analyzed "inverts" and transvestites. He found a large class of "professional" inverts, that is, male prostitutes, some them "pseudo-inverts," and urged the official surveillance and policing of sexual deviants, who were potential auxiliaries to crime in the subcriminal world, the "low-life" of Buenos Aires.[36] Veyga wrote in 1904 that "the world of the faggots [maricas] is so intimately linked with that of the petty criminals [lunfardos] and prostitutes, that one could say it is a part of both."[37] Homosexuality was not, however, only a weakness of low class. In his acclaimed 1908 book Mala vida en Buenos Aires, Eusebio Gómez said, "The group of existing homosexuals in Buenos Aires is numerous"; there were many opportunities to meet and hold dances; and there were many homosexuals in boarding schools and in the upper classes, especially among the youth. Gómez described male homosexuality as "uranism" and "pederasty" and lesbianism as "sapphism" or "tribadism." Was the cause hermaphroditism in utero? Or was it neurological, both acquired and congenital? Science had not "conclusively established the cause of this repugnant aberration." If the causes were not clear, "whatever the solution we will arrive at to explain this monstrous aberration, the individuals affected by it should be included, necessarily, in the picture of the 'mala vida' [lowlife]," of "defective morality."[38]

The Urban Male Criminal: Indolence, Regressive Heredity, and Alcoholism

Despite the medical field's fixation on female pathology and sexual anomalies, state officials could not avoid the evidence, based on turn-of-the-century police statistics, prison populations, and courtroom convictions, that the vast majority of crimes, between 82 and 97 percent, were committed by men.[39] Criminologists defined deviant women by their neglect of their domestic roles, but they judged men's behavior in terms of their public roles—working, voting, and creating the products of national culture. Of course, women had few public roles and no political rights and hence could only be measured by their private lives. Criminologists considered the most disturbing qualities of the abnormal man, by contrast, a lack of work ethic, an indolent nature, and an inability to provide for his family. The work ethic was the principal marker of successful masculinity.

Belisario Montero, an Argentine diplomat living in Brussels and self-described expert on the poor, described such men as "disgraceful ones." "There exists in all the great cities a considerable and relative quantity of these 'disgraceful ones,'" Montero wrote. "After having spent many months in poorhouses, and being considered hopeless cases, they ramble through the streets, dragging their rags, misery, vices and shameful lazy customs. They are not exactly criminals, because they lack the strength and energy to commit a crime. What predominates in them is above and beyond all the moral sickness of laziness, which leads them to a life more vegetative than animal. They are parasites, who, descending step by step, arrive at total moral abjection, the saddest physical decadence."[40] This negation of men's normal role could weaken the nation since the ability and willingness of men to work was closely tied to the nation's progress and modernization.

Crime was often a substitute for men without the proper work ethic. In 1888 criminologist Luis Drago wrote, "the word *to steal* is synonymous with *to work* in the community of thieves."[41] Cornelio Moyano Gacitúa's widely cited *La delincuencia argentina* (Argentine crime) devoted an entire chapter to the work ethic: "To look at an individual by his profession is to say what his social situation is, his education and literacy, his social means and his moral level."[42] Crime itself could become a "profession." Thus, criminologists Francisco de Veyga and Juan C. Córdoba identified the "professional delin-quent": "One cannot say anything of him but that he is a degenerate, and an intellectual degenerate." The criminal's profession was at least a choice; a man with no respectable occupation was by definition a degenerate.[43]

Legal medical reports submitted as courtroom testimony and reprinted in scientific publications reflected the prevalent gender stereotypes. In an 1898 trial of a man accused of the attempted murder of his wife, medicolegal experts offered as evidence for acquittal testimony that he was an exem-plary husband in the public sense: "He has always been a man of honest habits, industrious, and well regarded by those who know him."[44] But crim-inologists frequently described the male criminal, as they did male homo-sexuals, in feminine terms. Both fascinated and repelled by prison culture (for example, slang, codes, social networks, simulated madness), criminolo-gists were scandalized by the routine practice of sodomy in the all-male prison world, a passive sexual role the male criminal shared with women. Male criminals, the experts believed, shared atavistic traits with women

and the "lower" races and possessed an inborn propensity to the barbaric primitivism that the Italian criminologist Cesare Lombroso believed
women had by "nature." Male criminals, like women in general, were considered inferior in their grasp of culture, intelligence, and language. Their
use of slang was proof of this deficiency. The difference was that atavism in
men was expressed in criminal activity; in the eyes of criminologists, most
women were incapable of violence and cunning.[45]

Not economic need but immoral sensibility led to crime. Dellepiane cited
a police report that claimed "the idea of robbery as a necessity completely
disappears in the presence of the country's prosperity and her abundant
means of living. Poverty only exists in vicious people." Dellepiane constructed an image of the criminal as inherently urban, a view reflected in
many elite writings of the time. While he granted that country-dwellers had
vices, too, especially alcoholism and gambling, he considered rural Argentines morally superior. Crime and suicide rates were higher in the cities. The
male nature of crime was made more distinctive by the high proportion of
foreign men in Buenos Aires. Immigration had brought more men than
women to Argentina, and there was a sense that Argentine women, as well as
the larger society, were vulnerable to the bad habits and primitive nature of
men from abroad. Most criminologists toted up crimes in the city, where the
majority of immigrants settled, while virtually ignoring rural crime. The
city's wealth and riches, Dellepiane argued, attracted a dishonest type of
European, "creating a current of criminal immigrants that is sufficient to explain the increase in crime in the capital during the years 1888 and 1889."[46]
Argentina's liberal immigration policy and outdated criminal justice practices, according to Luis Drago, had attracted deviant types who threatened
to turn the country into an "obligatory field of action for the criminals
thrown out of Europe by the incessant persecution of the police and the
unquenchable severity of the judges." While Argentina did need "honest
labor," Drago warned that "precautions" must be established "against the
undifferentiated mass of adventurers and criminals, who, mixed in the
migratory current, multiply day by day the number of toxic activities."[47]

Nevertheless, Dellepiane praised Argentina's white European composition as racially "superior" to the rest of the region, but he blamed the
foreign-born population for posing great criminal danger to the country's
fragile civilization. He attributed its assaults on public stability to an undesirable European cultural tradition: "The bands or groups of delinquents,
criminal slang, the many methods of robbery or swindling that one observes in the Argentine Republic, offer great analogies with the similar

institutions of the Old World. The European pickpocket is perfectly represented in the Argentine *punguista*." In his second book, *El idioma del delito*
(The language of crime), Dellepiane identified a distinct criminal immigrant culture and language. He created a "Thieves' Slang to Spanish Dic-
tionary" with hundreds of words from the "criminal vocabulary," a special
language that set them apart from "decent people."[48]

Dellepiane's interest in slang was reflected in European studies at the
time. Lombroso, for example, traced thieves' slang to atavism and criminals
as biological throwbacks with "numerous points of contact" to the old
languages of primitive tribes. A product of "regressive heredity," criminals
had "devolved" into a state of barbarism. Dellepiane modified Lombroso's
idea: immigrants' slang was an "imitation," a bastardization of "ordinary
language" with two particularly disturbing aspects, a tendency to "spread"
to the rest of the population and an ability to conceal. Like eroticism and
pornography, with their graphic and picturesque nature, slang could seep
into the language of the general population through the "imitative tendency" of the lower classes. Dellepiane cited prisoners' coded language and
secret communications from cell to cell. "Criminal literature" resembled
that of "decadents"; the literatures of both "manifest a clear and obvious
common neurosis, of an identical mental imbalance."[49] Criminal language
in Buenos Aires had a predominance of foreign words, which made it, in the
words of Luis Drago, even more "capricious and inexplicable."[50]

Argentine criminologists were convinced that the immigrant element
contributed to one of the most distressing aspects of modern crime: *simulación* or malingering, the feigning of mental illness, found especially in a
specific type of con man or trickster. Ingenieros, who pioneered scientific
observations and studies of "*simuladors*," said to be running rampant in
Argentina, tested his theories in a 1910 report on Alejandro Puglia, an Italian
immigrant, murderer, cheat, and suspected malingerer, who had served
prison time in his home country. Puglia had displayed a number of psychological symptoms when subjected to numerous psychological examinations
in Italy and was diagnosed as "very unbalanced," with "mental degeneration." After immigrating to Argentina, he was arrested a number of times
for passing fake money, then for assault, and after a staged suicide attempt
in prison he was dispatched to the infirmary, to the mental asylum, then
back to prison. There Puglia learned that his sentence had been set at ten
years, whereupon signs of mental disturbance reappeared.[51]

The Criminology Institute physicians investigated Puglia's heredity and
uncovered alcoholism on the father's side and mental illness in three broth-

ers. They found "numerous physical stigmas of degeneration," including tattoos and alcoholism; they characterized his intelligence as "average" and his personality as "temperamental, violent, and impulsive"; and they noted a history of fraternizing with suspicious characters. The final diagnosis read, "Mental degeneration with episodic agitation and impulsive tendencies against persons, accompanied by a deficiency in moral sense and aggravated by chronic intermittent alcoholism." Could Puglia be considered a "normal man"? Clearly not, they said, but he was not necessarily mad. He had "profound mental anomalies," antisocial conduct "in a very dangerous form of inadaptation," and a mental condition that was chronic and incurable. He was both criminally insane and a malingerer—a "super-simulator [sobresimulador]."[52]

Building on Ingenieros's work, Francisco de Veyga argued that malingering was sometimes hereditary, sometimes environmental in its cause. "In many cases degeneration is congenital, finding itself in the ancestors of thieves, neuropaths, lunatics, alcoholics, arthritics, and sometimes in criminals. Other times degeneration is acquired, the product of pernicious influences in the social environment, including poverty, alcoholism, simulation, surmenage [mental exhaustion]."[53] Veyga examined a "known thief," called "el Ganzo," a Frenchman and chronic alcoholic with numerous physical and mental ailments, and pronounced him an "inferior anthropological type, degenerate." The shape of his hand exhibited traits that "some authors have described in professional thieves." After numerous arrests, el Ganzo was sent to the penitentiary's Lunatic Observation Room, where psychiatrists showed him off to students of legal medicine as a particularly interesting case of forensic psychopathology, a combination of criminality and insanity. They prescribed purgatives, bed rest, and a regime of "therapeutic hygiene." After he seemed to improve, he was dispatched to the mental asylum for long-term treatment. Physicians there suspected him of malingering because when they asked him simple questions, he answered absurdly and contradicted himself. He was discharged a few days later, bragging to a psychiatry professor to whom he was exhibited again that he had faked his symptoms in the observation room, tricking students and medicolegal experts. The final diagnosis was "simulation of madness by a degenerate professional criminal, convalescing from a temporary episode of depression produced by alcoholic intoxication."[54]

Veyga, in an article on "The Simulation of Crime," described malingering as logical to the criminal's struggle to survive: "He enters, in effect, in the plan of all crimes, especially of premeditated crimes . . . to avoid punishment." Malingerers, most often the young, falsified a criminal act to further their reputation, to hurt someone, or simply as a result of their brain chemistry. Real and simulated criminality were similar, Veyga wrote:

> This whole group of deeds which realizes itself under the mask of simulation contains in it something which is real; that is the germ of criminal will in which it originates. If this germ has not matured, it is not because of a lack of decision, but because the result intended by the act was obtained more rapidly and effectively by imagining it than executing it. Each time that one of these acts is carried out, there is the possibility of a criminal act. . . . In a word, each faked crime . . . encompasses the possibility of real crime which has a higher probability of converting itself into a deed than in remaining as it is.

In a sense it did not matter whether the criminal was true or dissembling; what was important was the social impact of the deception. In either case, such criminals presented a danger to society.[55]

"A Foreign and Hostile Horde": The Crowd

In the rationale of Argentine social pathologists it was only a short step from the degenerate immigrant individual to the degenerate immigrant mob. Working-class neighborhoods were suspect, in any case, because of their mushrooming growth, their dirt and disorder, their cheap bordellos, and their hordes of idle men. Criminologists and Argentine elites regarded urban crowds as especially alarming, a threat to become the modern equivalent of the violent and "uncultured" rural masses who had followed the rural caudillos in the earlier part of the nineteenth century. Those rural people of Argentina had seemed to them uncivilized, irrational, and even hysterical, just as the present foreign urban crowd was barbaric, irrational, passionate, and susceptible to suggestion.

José María Ramos Mejía announced clear differences between city and country crowds in drinking habits, class structure, race, and even shoe size and beard length.[56] Rural mobs were unintelligent, uncultured, "backward," and misguided in their critical support of local caudillos. City crowds, described in feminine metaphors despite the fact that most were men, were "impressionable and fickle like passionate women, purely unconscious; fiery, but short-lived; lovers above all of violence, of bright colors, of loud music, of tall and beautiful men; because the crowd is sensual, rash, and full of lust to satisfy its senses. It does not reason, it feels. It is unintelligent, reasons poorly, but has a large and deformed imagination."[57] The crowd was described as "hysterical," a mirror of the image of the hysterical woman in criminology: morally weak, fickle, suggestible, and shifty.

Social pathologists made the crowd a special subject of criminological investigation, turning in studies and, eventually, proposals for state control of the masses to the infant field of crowd psychology then emerging in Europe. The French physician Gustave Le Bon, in the 1880s, had explained social and political mass behavior like the widespread May Day demonstrations, anarchist terrorism, strikes, and violent demonstrations as psychologically determined. Le Bon contributed to a racial typology that was soon to spawn a branch of sociology that identified certain kinds of people as "barbaric," as atavistic throwbacks to an earlier, more chaotic time, and yet simultaneously as a symptom of the worst aspects of modernity.[58] "The Soul of the Crowd," an article in the 1905 *Archivos*, captured the tone of popular descriptions of the crowd in Europe and in Argentina: "The dynamic of suggestion, of imitation, of ignorant pretending, always incites more vigorously in the crowd. The state of affective excitement is more regular, more intense, more stimulated. It turns into collective agitation and, finally, exhausts itself in a state of delirium."[59]

In Argentina novelists, psychologists, animal behaviorists, criminal anthropologists, and historians relied on the methods of existing sciences, including hypnotism, pathology, and evolutionary theory, to portray the crowd in class and gender terms. An 1899 university textbook on the new social sciences, *Introducción general al estudio de las ciencias sociales argentinas* (General introduction to the study of Argentine social sciences) by Juan Agustín García, claimed, "Anonymous heterogeneous groups . . . are in all ways morally and intellectually inferior to the individuals that compose them. It is likely that the principal factor of this degeneration is the sentiment of *irresponsibility* that surges in the breast of all crowds with extraordinary force of action, and that explains its crimes and abuses."[60]

José María Ramos Mejía's *Las multitudes argentinas* (Argentine crowds) was the first application of European crowd psychology to a study of the Argentine masses. Ramos Mejía had originally intended this book to introduce a study of the nineteenth-century dictator Juan Manuel de Rosas as a man who had emerged from the "masses." But the author extended the study to warn against the mobs of Buenos Aires as more "primitive," and a more grave danger, than the crowds of Europe: "If modern man in European societies, who alone is cultured and moderate, demonstrates such barbarism when part of a crowd, just imagine how American throngs would be, as they are formed from a more violent and instinctive element, and are more subject to the enthusiasm and heroism of primitive beings."[61] In the social science textbook García wrote, "The Saxon crowd is different from the Latin, the Parisian throng [different from the] Spanish, English, or Argentine. The evolution of our crowds, their sentiments, ideas, acts, and conditions of their leaders, [and] the ways in which they suggest, are the principal questions of a national psychology."[62]

The crowd was thought to harbor criminals, terrorists, and their potential followers, but most perilous were members of an increasingly militant working class, who were often, the elite feared, led by anarchists. In the first ten years of the new century, Argentina experienced three major terrorist attacks and between 100 and 300 strikes per year, much of the turmoil the ordinary demonstrations of a democratic society. Yet to social scientists and government officials the unrest reflected a decline in decreasing public "morality" and escalating public danger.[63] General strikes and other proletarian events would incite violence and create conditions for greater social chaos. One author even included "worker rebellion" in his list of violent crimes.[64] And in 1903, when Buenos Aires city police commissioner José Gregorio Rossi, writing for an audience of scientists, decried urban mobs of "professional criminals," immigrants, and the poor, he equated criminality with foreign masses, which by then, in Buenos Aires, composed nearly half the population: "We have one serious factor which leads us to a strong growth in crime: it is the heterogeneity and continuous mobility of the city's population, due in large part to immigration."[65]

Cornelio Moyano Gacitúa's La delincuencia argentina (Argentine crime) agreed with European theorists that the "Latin race," in Argentina as in southern Europe, accounted for the highest crime rates. "Not one criminologist today," Moyano Gacitúa wrote, "assigns lesser criminality to the immigrant than corresponds to his nation of origin; on the contrary, they arrive at the categorical conclusion . . . that the criminal increases his harmful activity in other countries." Rossi, too, suggested that congregating foreigners were in large part responsible for rising crime, since due to immigration "we receive [in Argentina] entire groups of professional thieves."[66] The racial view of degeneration received in-depth treatment in a 1912 medical thesis by Abel Sonnenberg of the University of Buenos Aires. Religion, education, and customs could affect some types of degeneration, but hereditary factors like mental state, alcoholism, syphilis, tuberculosis, and birth anomalies influenced others. The faces and bodies of individuals showed sings of degeneration. Sonnenberg noted thirteen types of misshapen heads and numerous anomalies of the mouth, lips, teeth, ears, nose, and tongue. Referring to one of Cesare Lombroso's studies that had measured the ears of thieves, rapists, and murderers, Sonnenberg also reported a higher proportion of anomalies in degenerate persons. He drew from Argentine data, mostly culled from mental patients' and prisoners' medical reports. "Very numerous" cases of "psychophysical degeneration with impulsive stigmas of varying degrees of criminal and antisocial delinquency" could only be deciphered using "clinical criteria based on morbid psychiatric and psychological knowledge."[67]

Similarly, the first chapter of prison official Eusebio Gómez's *Mala vida en Buenos Aires* identified immigration and urbanization as causes of crime. He had included photographs of prostitutes, homosexuals, beggars, fortune tellers, and medical quacks and a variety of degenerate types of "professional thieves," homosexuals, and transvestites. Even José Ingenieros, an immigrant from Italy, wrote in a prologue to Gómez's work that Argentina's criminals constituted "a foreign and hostile horde." Ingenieros praised the book as a "realistic descriptor" of the "Buenos Aires lowlife" and its myriad "vices."[68] The *Archivos*'s Helvio Fernández, Ingenieros's successor as the journal's editor, praised *Mala vida en Buenos Aires* for identifying "characteristics" of Argentine crime, specifically of blood crimes, which were a consequence of the "quality" of immigration to their country.[69]

Crime theorists seldom blamed social and economic conditions for urban unrest. Miguel Lancelotti, a well-known lawyer and criminologist who worked with Pietro Gori and Ingenieros, did point to poverty and unemployment and the declining moral education of children as root causes of rising criminality, but his 1914 statistical study of crime as reflecting Buenos Aires's density of population saw "vice and immorality" in the big cities as providing "the biggest and most favorable conditions" for criminal auxiliaries like prostitutes, gamblers, and beggars.[70] But Lancelotti was virtually a lone voice in the general chorus of condemnation of the inferior and dangerous crowds of foreign elements.

Still, Argentine physicians and social scientists did not wholly accept the common European understanding of degeneration as a result of race mixture, and they differed also with their neighboring colleagues in Brazil, who saw descendants of African slavery as an urgent reason for racial hierarchy.[71] Although Argentine scientists considered degeneration in their country as having been largely imported by inferior southern Europeans, its effects could be mitigated with the right social environment. Elites pinned their hopes on assimilation. A "healthy" Argentina could produce a superior second and third generation. Ramos Mejía, in *Las multitudes argentinas*, managed to combine classic Lombrosian body measurement with a Lamarckian belief in environmental influences. "The first generation," he proclaimed, "is often deformed [*deforme*] and not good-looking until a certain age. . . . There are a certain proportion of pug noses, large ears, and thick lips: his morphology has not been modified by the chisel of culture. In the second [generation], one can already see the corrections that civilization begins to imprint. . . . The change in nutrition, the influence of the air, and the relative tranquility of the soul from easily obtained food and the supreme necessities of life, affect their transcendental influence."[72]

The Worst Type of Criminal:
The Anarchist of "Degenerate Lineage"

The most harmful aspect of the immigrant crowd, so the social scientists believed, was its tendency to harbor the anarchist, that self-proclaimed enemy of bourgeois society and the state. While worker unrest and anarchist terror had been a persistent threat since the 1890s, the year 1905 brought violent class conflict, including the unsuccessful 11 August assassination attempt on Argentine president Manuel Quintana. Argentine social scientists experienced near panic that their young nation was doomed to experiences similar to the tumultuous demonstrations and violent clashes then rocking Paris, London, and Barcelona. One criminologist, Ramón Pacheco, after listing more than a dozen terrorist attacks worldwide, expressed such anxiety and despair: "The multiple attacks, for the most part anarchist . . . call to the attention of all thinkers those acts which have painful repercussions for all humanity. . . . [That] civilization and progress open a path in spite of all obstacles, would make us believe that humanity could complete its evolution in peace, through the work of reason, and leave aside violence and the spilling of blood."[73]

It was widely believed that most anarchist and labor leaders in Argentina were Italians, Spaniards, or Jews who had escaped justice for misdeeds committed in their homelands. Elites imagined Argentina to be the final laboratory for these foreign troublemakers, at once welcoming these rebels and inadvertently providing fertile ground for their incendiary ideas. While the elite believed that both "good" as well as "bad" types of immigrants populated Buenos Aires's crowded neighborhoods, it warned of newcomers bearing controversial ideas, including doctrines of extreme social change or even destruction of the capitalist system on which Argentina depended. In

1907 Gómez wrote that the great number of immigrants in the city added to the "army of proletarians" and the subsequent "conflicts between capital and labor." Attributing "lowlife" to the labor movement, he portrayed workers' tactics as "saturated with hate and zeal to destroy."[74] Anarchist ideology and action were aimed at igniting the nation's most flammable conditions: housing, labor and political rights, and even women's rights.

In one of the first "scientific" studies of anarchism, published in 1897, Francisco de Veyga had associated anarchism with the crowd, though he pointed out that in spite of many similarities, the two should not be confused—anarchism was a derivative of the crowd and as such took on many of its attributes. Both were, however, primitive throwbacks, suggestible, superstitious or religious, and irrational. Like individual members of the crowd, the anarchist was susceptible by nature to criminal behavior: "Every anarchist today is predisposed to crime. All are equally fanatical and resolute, [and] do not need an inspiration for crime other than the example of those who have already fallen, whose painful but sublime footsteps they follow to their own sacrifice. To kill and destroy is perhaps an excuse to become a hero." Veyga did not see anarchists as true revolutionaries. They had "no faith, no altruism, nor revolutionary impulse, but only insanity or inborn criminality"; they were either mentally ill or ill willed; and the causes of their criminal acts were buried in their physical, mental, and "moral" states. Veyga dismissed the demands of anarchists and socialists to reform the living conditions of the poor or improve workers' lives as inconsequential to the true causes of crime; he looked instead to inborn causes. Recounting the history of attacks to support his thesis that anarchists could be classified as criminals, he concluded that certain historical figures could be understood in terms of modern criminological theories: "the clinical history of each criminal anarchist is exactly the same as that presented by a political criminal foreign to that creed." He named a number of notorious European anarchists of the day as "born criminals," "degenerates with political delirium," and in other Lombrosian terms.[75]

Discussing the attempted assassination of President Quintana, Veyga identified the attacker as a Spaniard with physiological stigma, from a "very humble family," with a sister who suffered either hysterical or epileptic attacks.[76] Veyga's in-depth 1906 medicolegal study of the assailant, Planas Virella, cited both biological and psychological "irregularities." "Spiritual anomalies occur frequently among assassins," he wrote. Veyga noted Lombroso's subdivision of anarchists into three types: those who use crime to act out their "perverse intentions"; the insane anarchist; and anarchists of "passion" or overexcitement. The anarchist was closer, to use Lombroso's widely adopted phrase, to the "born criminal." Veyga's physical and psychological examination of Planas Virella found "pronounced facial asymmetry . . . a malformation of the opposite half of the skull. One notes another flaw in the bones of the nose, giving this organ a childlike aspect. . . . Average brachycephaly [a deformation of the skull], resulting perhaps from his ethnic influence." Veyga also noted an "exaggerated reflexive sensibility, with a considerable energy in his emotional releases. From this same reactional exaggeration derives in him a much-pronounced frequency in impulsion, such that this tendency played a direct role in the decision to commit this crime."[77]

In 1909 Buenos Aires was again thrown into panic when a young anarchist, immigrant Simón Radowisky, assassinated the city's despised police chief, Ramón Falcón. Radowisky, who later became a national icon to many radical workers for his defiance of police and prison officials, had emigrated from Russia as a teenager, but he kept the facts of his life obscure, most likely to stymie investigations of his prior anarchist activity. Medical and psychological examiners sought to determine Radowisky's age, motives, and dangerousness. Radowisky represented the specter of unknown foreign dangers festering in the crowd, influencing it, supported by it, and potentially inciting it to overt rebellion. His act spurred new social control laws (see also chapter 10) and an outpouring of racialized discourse on Argentine anarchism. Newspapers and criminal justice records alike undertook intense scrutiny of his background, most notably his Jewishness, a fact that fed the public's imagination of the crowd's ethnic solidarity and class resentment.

Following Veyga's study, criminologists relied on biological theories of anarchist violence, focusing on the anarchist as a distinct criminal type and formulating a general psychological profile. In a 1910 study published in the *Archivos*, the noted Spanish criminologist Bernaldo de Quirós summarized current European and Argentinean theories and cited the principle (which he attributed to José Ingenieros) that anarchists' individual psychology (like everyone else's) stemmed from heredity, as well as from "the environment in which one lives and is educated." Anarchists were of "degenerate lineage, germinating under suggestions of hate and arriving at anarchy, sometimes by the biological law of mimicry, or other times from the desire for notoriety, the insane pride of infamy." They were pathological, for the most part, hypersensitive and delusional. Most anarchists committed their acts in a futile attempt to alleviate the "universal pain" to which they were acutely sensitive.[78]

Argentine social pathologists' efforts to determine the levels of severity of deviance and to find ways of marking, and hence segregating, dangerous individuals had resulted in the identification of types they believed undermined moral character and behavior. Scientists and government leaders linked a wide range of "degenerative" problems, from tensions raised by female employment to the culture clash introduced by immigration and explosive class conflict, directly to attacks on property, economic development, and thus national interests. Considering rural interior types more manageable and, therefore, believing Argentina's real problems were urban, they warned that urban crowds of immigrants were the most immediate menace. The scientists' diagnosis of the urban ills of society led to an extensive program of social engineering. By 1912 Argentine eugenicist José Angulo was exhorting "criminologists and sociologists, rulers and legislators" to join forces to "wage a merciless war against the degeneration that threatens the future of humanity."[79]

PART III *Prescriptions*

FIVE

Women Confined to Save the Future Nation

Home and Houses of Deposit

> The function of the home, [and] its proper action in the formation of moral
> ideas, is to develop them. . . . The child derives unawares his habits and ideas
> in great part from the models and stimuli that he receives in the home.
> —VICTOR PESENTI

A typical view of Argentina's gender roles appeared in one of the twentieth century's first books for girls, a three-volume set published in 1902 entitled *La niña argentina* (The Argentine girl) that was designed to "instruct and prepare the future Argentine ladies." The author, Rafael Fragueiro, assured the reader that the best works from "European and American writers and educators" had been selected for the "girl students of the Argentine Fatherland." A lesson on the family began, "God created the family." Written in large, bold letters for the convenience of young or recent readers, it described the ideal bourgeois family:

The family is composed of the mother, father, children, siblings, and relatives.

In the Christian family we also consider the servants members of the family.

One of the most beautiful sights there is, is that of a good family, respectful and united.

The father works, teaches, and protects his wife and children. The wife loves and respects her husband, raises and cares for the children, and takes care of the home.

The children love and respect their parents. They obey. They cheer up the house, and when the parents are old, they work for them, and help them with their needs.

It ended with the admonition that God would bless the families that achieved these ideals and that "all men would admire them and be influenced by their example."[1]

So, too, novels by Generation of 1880 figures, like Eugenio Cambaceres, portrayed the proper family environment, but they also depicted how it could go wrong in the lives of creoles, immigrants, and second-generation Argentines alike. Novelists warned of the consequences of improper behavior like adultery, child abandonment, excessive drinking, and gambling. Cambaceres's 1885 novel *Sin rumbo* (Aimless), for example, told a tale of rape, adultery, illegitimacy, and suicide. Antonio Argerich, in his 1884 novel *Inocentes o culpables?* (Innocent or guilty ones?), described a young man's visits to brothels and his suicide after contracting syphilis. In the 1899 novel *Irresponsable* (Irresponsible) by Manuel Podestá, a man's decline into mental illness was linked to alcoholism and his association with a prostitute.[2]

In biology, medicine, and the human sciences in Europe and North America, social theory was just starting to emphasize the importance of the family and the home in creating healthy families and children and to rethink the public and private roles of men and women. Central to these projects was an awareness that children, shaped by their home environment and especially by their mother, were a resource for the nation as its future citizens. The mother's moral education could prevent in her children homosexuality, alcoholism, crime, suicide, and mental illness, all products of pathological modernity and the noxious side effects of Argentina's attempts at "civilization." Mothers were, therefore, a central concern of Argentina's scientific state. Intervention was called for in what had traditionally been a private arena guided by the Catholic Church.

The era of increased government surveillance of the private sphere brought multiple inspections, studies, censuses, and reports. Social reformers, turning a clinical eye to the urban life of the poor, started at the most private sphere of all, the bedroom, and then moved beyond the living areas out to the larger neighborhood and the street, where men (and prostitutes) were the principal concerns. The reformers proposed quite different ways of dealing with, and controlling, problematic behaviors in men and women. Their solution to women's wayward behavior was to confine them to the domestic sphere where they could be controlled. Although the Catholic Church still held sway over many of the issues of private behavior, in this period a struggle ensued between the church and those physicians and hygienists who supported secularization and who wanted to replace the church's hold over private lives with ideas and techniques of their own, for example, home inspections, clinical visits, and scientific training for mothers. In some areas the medical professionals prevailed. In others, the professions backed off and left women in the hands of the church.

Through state-sponsored studies of women's maladies, psychiatrists and social reformers began to intrude into women's lives in new ways. Some women were required to consult with physicians at the mental health clinic of the city police—state employees—about intensely private domestic matters. Other women were inspected and assessed by state medical officials in their homes. Still others were measured, examined, and analyzed in the city's mental hospitals and syphilis clinics. The government and social pathologists feared that any disease would put in jeopardy the "natural order" of the domestic, reproductive family unit. A physician's intervention was understood not just to alleviate a woman's individual suffering but also to heal the breach in the social fabric created by degenerate children. The goal was to make the family healthier. Doctors examined the physical body for signs of degeneration and moral contagion, and they formulated prescriptions to ameliorate symptoms and to control the family to prevent further erosion. The social pathologists explained that working-class families in Buenos Aires's burgeoning slums were especially vulnerable. They pointed out that inhabitants of the slums and *conventillos*, the crowded tenements teeming with children, faced numerous threats to the well-being of the family, among them venereal disease, crime, alcoholism, even a mother's work outside the home. Science also sought to control abnormal sexual behavior, including homosexuality, to prevent its spread to larger society.

Physicians, hygienists, and psychiatrists supported by the state advocated secular control of government agencies. While claiming to be detached from religious ideals of morality, they held views of vice and virtue that were remarkably similar to those of religious institutions. It was a moment of transition in Argentina, an opportunity to turn away from traditional practices and church control. Science and medicine hoped to take over the effort of controlling, guiding, and educating women in the home, but when it came to women's roles, the new liberal views were remarkably similar to those of tradition and less in tune with a secular Enlightenment perspective. Both church officials and state scientists saw the family as the foundation of a peaceful, orderly society. Both demanded that women submit to male authority. Both ascribed to motherhood the "natural" central role for women in the family. Traditional gender roles in the family thus

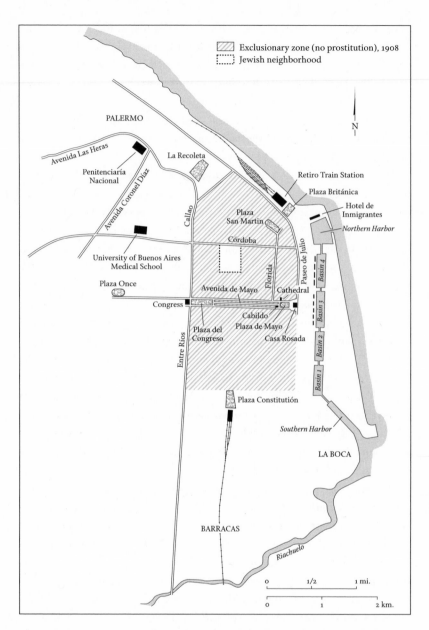

Legend:
- Exclusionary zone (no prostitution), 1908
- Jewish neighborhood

PALERMO

Avenida Las Heras

Penitenciaría Nacional

La Recoleta

Avenida Coronel Diaz

Callao

Plaza San Martín

Córdoba

University of Buenos Aires Medical School

Plaza Once

Avenida de Mayo

Congress

Florida

Cathedral

Cabildo

Plaza de Mayo

Casa Rosada

Plaza del Congreso

Plaza del Congreso

Entre Ríos

Retiro Train Station

Plaza Británica

Hotel de Inmigrantes

Northern Harbor

Paseo de Julio

Basin 4

Basin 3

Basin 2

Basin 1

Plaza Constitutión

Southern Harbor

LA BOCA

BARRACAS

Riachuelo

N

0   1/2   1 mi.

0   1   2 km.

Map 2. Buenos Aires, Waterfront and Surrounding Barrios, circa 1910. (Based on James R. Scobie, *Buenos Aires: Plaza to Suburb, 1870–1910*, 1974, and Donna Guy, *Sex and Danger in Buenos Aires: Prostitution, Family, and Nation in Argentina*, 1990)

remained largely untouched. Indeed, women who transgressed were left in the hands of the church when in need of discipline. Science had not changed fundamental assumptions about gender or roles for women.

The "Pride of the Kitchen, Bedroom, and Parlor" but Prone to Hysteria

Traditionally, as in most of Latin America, the Catholic Church in Argentina, which was more than 95 percent Catholic, had defined the expected roles and private behavior of women and men. The church had historically considered itself the key social authority in the maintenance of social order and cultural integrity. But in the 1880s, with anticlericalism and secularization in Argentina at a peak, the liberal modernizers assumed new authority for judging and shaping private behavior. The state now took control of education, birth and marriage registries, and cemeteries. New laws took over power from the church, providing for, in 1884, civil registration of births and marriages, civil marriage in 1888, and then secular administration of cemeteries, in part to prevent contagion from unsafely interred bodies. Education reform, even more sweeping, had created a national system of public primary education in 1884, its goal to assimilate the masses of immigrant children. The church was no longer permitted to teach religion in the public schools, though this remained a controversial issue.[3]

Secular officials similarly sought to dilute the church's hold on the domestic sphere, but in the end they agreed with religious leaders on the fundamental roles for women. Modernization schemes had a limited effect on women's lives. Although the social pathologists' ideal woman was intelligent and knowledgeable, her skills were to be applied to the task of raising children and to running an orderly, moral, and sanitary household. A woman was to be devoted above all to the comfort of her husband and to the raising of healthy children. A veneer of bourgeois values and middle-class goals, largely unattainable, was imposed on working women as well. (Feminists made similar class judgments, as in 1901 when Elvira López stated that "the fewer proletarian women there are, the stronger our race and our social morality."[4]) Victor Mercante, an educator who supported education for women, did so on the grounds that educated women would contribute to the home and, thus, to the nation's progress. "The mother," he wrote, "devoted to the care of her children[,] bound to those of her home," would excel as "an ordered soul, exalted administrator, the pride of the kitchen, bedroom, and parlor."[5]

Raising a family as "an ordered soul" in a *conventillo* or other type of crowded tenement building, downtown or in the outlying suburbs, was demanding and exhausting. In the early years of the twentieth century, about four-fifths of the city's working population lived in these tenements, with most families—on average, five to seven people—crowding into one room approximately twelve by twelve feet. Women cooked on small charcoal stoves in the hallway outside the room. During children's school hours mothers did piecework for hire, sitting in the common hallway. There were many orphaned and abandoned children, and widows made up about 15 percent of the population. About 14 percent of all births were illegitimate, which left families economically vulnerable. Large numbers of women with children and without access to a man's wages were forced to work for money, but their job options were limited to low wage labor in small textile factories, laundry and piecework at home, or prostitution.[6]

In many families, mothers (and even children) worked outside of the home to support the family's upkeep. In 1869 about 39 percent of the city's women worked for wages, though the rate steadily declined as the new century approached.[7] The 1869 census's tabulation of professions found women working principally as seamstresses, laundresses, weavers, ironers, cigarette makers, and cooks. The census took a separate look at prostitution, noting that "only 361 individuals, of both sexes," had reported this profession. (The census taker estimated the actual number to be at least ten times higher, "half of the adult female population," women with "difficult and precarious" lives who suffered under the "uncertainty of daily sustenance.") Wages, which had drawn many immigrants to Argentina, were depressed in this period, and prices were high. The newspaper *La Prensa* reported in a series on labor in 1901 that the highest monthly earnings of a male day laborer, seventy pesos, were thirty pesos short of what was needed for his family. The census's commentary saw female immigrants' labor as contributing to the "infinitely fertile expansion of the tendencies and industrial aptitude of the nation."[8]

Women were held responsible for the domestic sphere, including household chores, reproduction, and sexual service to their husbands. In a 1910 medicolegal report, the examining physicians of one "M.U." complained that she "does not feel the attractions of family and society."[9] Physician and national legislator Lucas Ayarragaray urged women to keep themselves attractive, lest their men lose sexual interest in them.[10] A lone feminist voice took a different view on marriage. An 1896 article in *La Voz de la Mujer* (The woman's voice) commented on a case mentioned in *La Prensa* that de-

scribed a Manuela Bermudez from the town of La Pampa who had murdered her husband and family. "Cause of crime. The unbearable life that
Cutiellos gave to his wife. Commentary of *La Prensa*: That all the rigor of the
law falls in the face of this infamous wife and heartless mother."[11]

Like the Catholic Church, physicians and secular reformers were especially concerned with control of women's fertility and sexuality. The
church's view on these matters was widely accepted across class and ideology, and among the scientists themselves. In the population at large (and
even among most feminists, anarchist women being the exception), traditional Spanish and Catholic ideas about women's purity, based on virginity, still held sway. Nonetheless, some physicians, psychiatrists, and hygienists had begun to speak and write candidly about sexual behavior. They
promoted sexual education, and some even advocated sexual pleasure for
women as a way to increase the birthrate and healthy mothering instincts,
not to mention marital stability and the husband's satisfaction. Birth control measures, however, were to be denied to women and men alike.[12] Abortion was illegal, immoral, and out of the question. Divorce was unhealthy
and destabilizing. Old ideas merged with new scientific approaches, as
in Elvira Rawson de Dellepiane's 1892 medical thesis on "women and hygiene," in which she advocated the introduction of sex education but cited
healthy and moral womanhood as the primary goal. She stressed the importance of sexuality within the context of a "healthy" marriage and family.[13]

Psychiatrists employed by the Lunatic Observation Service of the Buenos Aires city police, open in 1899, received, studied, and treated dozens of
mentally disturbed individuals, men and women alike, and provided new
evidence of women's disruptive effects on marriage and the family. For
example, José Ingenieros examined a suicidal young woman brought to the
clinic by her distraught husband. Writing the case up in the *Archivos de
Psiquiatría, Criminología, y Ciencias Afines* (Archives of psychiatry, criminology, and related sciences), Ingenieros reported the root cause of her mental
disturbance: "The young woman . . . confessed that during two years of
marriage she had not once experienced any sexual feelings." This sorry state
of affairs, according to Ingenieros, was the result of her "complete sexual
ignorance," which in turn resulted from her repulsion at the sexual act,
making her feel "inferior for conjugal life." According to the patient, these
feelings had led her to attempt suicide; she would rather die than be abandoned by or humiliated in front of her husband.[14]

In 1909 the *Archivos* published the proceedings of a civil trial under the
heading, "Sexual Impotence as a Cause of Divorce." A man had taken his

wife to court, demanding an annulment because of her lack of interest in sexual relations and motherhood. He accused his wife of neglecting her conjugal duties and pleaded, "My wife, then, is sexually impotent. Her sexual organs do not impede her from performing, and I believe that they are well formed, but what is indubitable and evident is that the cause of her impotence does not reside in her organs, but rather originates in a mental cause."[15] Even the sex lives of "normal" married women were subject to critical review by social pathologists. Lucas Ayarragaray, physician at the National Women's Asylum and later national representative in Congress, claimed that "psychological impotence . . . does not only depend on the man, but is sometimes provoked by the conduct, lack of attraction, etc., of the woman, who temporarily kills the sexual [*genital*] abilities of the man."[16]

The husband who took his wife to court testified that "the character of my wife corresponds to all that which is known in science about the affective traits presented by hysterics. . . . Hysterical women are, then, by their very spirit, *duplicitous* types who *lie* and *simulate*."[17] Physicians who acted as expert witnesses in court cases presented hysteria, then seen as an epidemic among women, as the antithesis of marriage and maternity. Hysteria, according to psychiatrists, was one of the biggest threats to healthy marriage. Ingenieros commented on a typical case of hysteria from 1910 that not only was his patient suffering terribly, but she also was unable to fulfill her biological destiny because of her lack of sexual knowledge. Ingenieros recommended that the husband pay closer attention to his wife during the sexual act, allowing her to build up "sexual emotion." After just a few weeks, the doctor noted with satisfaction, "this simple sexual education permitted the reconstitution of a home" that had been about to collapse. The messages implicit in Ingenieros's diagnosis are surprising, and even contradictory. On the one hand, he exhibited an open-minded, even liberationist, attitude toward women as sexual beings, which was unusual in a time characterized by a cultural double standard in which men were permitted to be sexually active and women were not. Ingenieros and others who saw themselves as modern and progressive were offering alternate interpretations of private behavior, solidly informed by the evolutionary theories favored by the young generation of scientists—most people rejected the application of those ideas to the sexual behavior of women. On the other hand, even Ingenieros seemed to reduce his patient's mental suffering to her sexuality, specifically, her inability to fill her sexual obligations to her husband. At the turn of the century, numerous clinical studies and published medical research on "women's diseases" conveyed similar views.[18]

As a remedy Ingenieros proposed hypnosis, a popular French cure for hysteria. A large portion of his 1904 book *Histeria y sugestión* (Hysteria and suggestion) was devoted to hypnotic techniques and their social applications. Hypnosis, or the "state of suggestibility," had acquired some legitimacy thanks to the influence of French neurologist Jean-Marie Charcot. The hypnotic trance's sequence of bodily states could be scientifically measured. Hypnotic states were distinct from neurological conditions such as catalepsy and sleepwalking, and only some individuals were hypnotizable, or suggestible. Hysterics were considered prime candidates for hypnotic therapy because doctors believed that women, who comprised the majority of such cases, were suggestible, gullible, and weaker willed, and therefore were disposed to hypnosis. Theorists assumed that suggestible individuals displayed a lack of agency or, in Ingenieros's words, an "ailing volition." Ingenieros believed that hysterical and hypnotic states had similar attributes but that the two should not be confused, "because hysteria is born from self suggestion and hypnotism through outside suggestion. The hysteric is active, the hypnotic is passive." He argued that suggestible subjects "cannot govern themselves any more; [they] lack proper will, an independent ego, and normal psychic activity." A seventeen-year-old victim of hysteria, caused, it was assumed, by masturbation, responded well to the new psychiatric methods of treatment. The experts had concluded that her relieving her sexual urge through "unnatural" means was inextricably linked to the signs of her disease. Ingenieros recommended a two-part treatment: for her hysteria, laxatives and warm water therapy; for her laughing fits, suggestive therapy and hypnosis. After a few months of successful therapy, doctors trained her mother to continue limited treatments at home (see also chapter 4).[19]

Although for conditions like hysteria, most physicians recommended laxatives and warm water or hypnosis, and in extreme cases admission to one of the city's new specialized mental asylums, most sent women who displayed abnormal behavior home for domestic therapy.[20] Motherhood itself was considered a prescription for certain types of mental disturbance. One researcher concluded in his study of suicide that "that which preserves the mother best from suicide is maternity."[21] Even abused women were expected to remain in the home. In a 1907 case of domestic abuse, "A.O." took the unusual step of appealing to the public assistance office to intervene. Her husband had repeatedly raped her, causing injury. The authorities ordered both individuals to be examined by the court's medicolegal experts, and the specialists concluded that "A.O." did not exhibit signs of abuse and that "the husband's penis is normal and is not exaggerated in size or length." The judge, Ernesto Quesada, a frequent writer on criminological topics, ordered the woman to return to her "marital home" within forty-eight hours, citing Article 210 of the civil code: "the wife is obligated to live with her husband . . . . [I]f she fails to do so, anyone may order the necessary judicial means." Quesada said the medicolegal evidence pointed to the "categorical conclusion that the sexual conditions of both husband and wife are normal, and no one suspects that intercourse would be harmful to the life or health of [A.O.]" The title of this case in the reprint was "On the Fulfillment of Matrimonial Duties."[22]

Domesticity was itself a cure for women who deviated from ideal behavior. Criminologist Antonio Dellepiane, echoing the influential Italian school theorists Cesare Lombroso and Guilliermo Ferrero, as well as traditional Argentine views, proposed as the solution to lapses of and deficiencies in female morality the "moralizing force of matrimony." Married women, Dellepiane said, had the lowest crime rates and complained of falling ill less often. Marriage had an even more salutary effect on women than on men, due to "the development of maternal sentiment and the obligations which that brings." The greater religious tendencies and altruism of women diminished the likelihood of criminal behavior: "The state of matrimony is a form of crime prevention." Thanks to their well-defined roles, society "makes women more moral than men."[23]

Social pathologists, as staunch secularists, opposed church domination of marriage and family life, at least in theory, and some even supported the right to divorce, at the time still illegal in Argentina. A healthy family could only exist, proponents of legal divorce explained, if husband and wife were happy with one another. A 1903 article in the *Archivos* declared that "the question of divorce is not only of interest to the couple for whom a common life is impossible: it also pertains to their families, friends, relatives, and has repercussions for the customs of the whole society."[24] Liberals had won civil marriage in 1888; the secularization of marriage had been one of the most important issues to reformers who aimed to strip the church of its power in Argentina, and one of the first addressed by the liberal oligarchy after they took over in 1880. Civil marriage had been implemented to the dismay of the church and traditionalists, but the church was more successful in defeating proposals aimed at legalizing divorce. At the turn of the century, divorce was highly unpopular among the Argentine population, though it was supported by many liberals, socialists, and anarchists on anticlerical grounds, as well as out of concern for women's equality in marriage.

Many opponents to divorce often saw it as dangerous to women since they were vulnerable to abandonment and had few means of support without a man. Others were disturbed by the implication of marital equality, especially in instances of adultery, lest it appear that the state condoned immoral behavior in women. Above all, there was the concern that divorce would destabilize the family and lower the nation's birthrate. While both partners in the marriage were required to be both moral and loyal to the union, women paid the highest price for betraying this family ideal. Women were in practice held to a higher standard.

Wayward Wives, Women on Deposit, and Feminist Responses

Women who failed at their "natural-born" domestic roles faced a range of consequences. In extreme cases, though not infrequently, husbands could sue their wives for "lack of female role" and have them placed on "deposit" in specialized institutions. "Houses of deposit" for wayward wives who disobeyed their husbands (or fathers, if unmarried) or otherwise put the family honor in jeopardy, had no male equivalent. Control of these houses of deposit, a vestige of colonial times, fell to the church and to the state, which shared the organization, oversight, and operations. Women of all social classes could be deposited not only at the request of family members but at the request of police, judges, or even employers. A civil judge had the final say in ordering sequestration. Most women served several months for one of a number of offenses, ranging from "bad temperament" to infanticide, abortion, and adultery, at one of three institutions designated for women's confinement, which had been set up by the church or the quasi-governmental ladies' charity organization, the Beneficent Society. For example, between 1862 and 1905, about 100 women were sent to the Good Shepherd.[25] Minors, younger women, and those professing great religious sentiment were deposited at the Spiritual House or the Good Shepherd; most others were sentenced to the Women's Hospital, where they provided unpaid labor to the institution and slept under patients' beds. Even the administrators of the crowded and filthy wards protested the conditions under which deposited women served. Wives opposed depositing by resisting their own incarceration and initiating lawsuits against their husbands.

While a large-scale movement to end the practice did not emerge, Argentina's first feminists began in the 1890s to advocate for legal equality for women. As women in growing numbers joined the work force, a few demanded in ever louder voices a stake in the traditionally male arenas of education and politics, in the hope of changing the economic and political status quo. Such views appeared unimaginably unrealistic to most in Argentina, where beliefs about women's fixed roles were supported by both religion and science. Two strands of feminism, liberal and socialist, joined the public debate on legal reforms, with liberal feminists directing their energies toward enhancing the rights of married women, promoting equality of the sexes, and establishing property rights of women. Liberal feminists, who counted among themselves a number of the nation's first female physicians, such as the patrician Elvira Rawson de Dellepiane, focused on legal reforms to give women financial autonomy and political reform, especially suffrage. Socialist feminists concentrated on working women and their health and standards of living, advocating welfare reform and objecting to the prevalent sexual double standard. They were also against reproductive control. After 1900 socialists stood alone in supporting expanded civil rights, equal labor opportunities, and suffrage for women.

Both liberal and socialist feminists challenged male superiority in the law, yet most women, while struggling to attain more rights in and outside the family, hesitated to give up the special status they held through their capacity for motherhood. As a result, women, like men, were disturbed by prostitution, abortion, and exhibitions of uncontrolled sexuality in the mentally ill, which all seemed to them to risk women's maternity, and most accepted a separate sphere, though insisting on women's rights within it. Most social reformers, including feminists, who supported equality for women in the public sphere did not address inequalities in the home, and dualistic gender divisions were taken as natural, except by socialist women. Most believed that women should be better educated, but they did not question the primacy of male dominance in the home. The majority of feminists agreed with more traditional policymakers that women and men were fundamentally different and that women's unique feminine traits ought to be preserved. They praised the sanctity and holiness of motherhood, along with women's supposedly innate talents for nurture and sensitivity. They saw complementary roles for women and men, though in the context of legal equality. Maria Abella de Ramírez, founder of the National Feminist League of La Plata, summed up this view in 1919: "What is a woman feminist? Rumor has it that she knows nothing about the poetry of the home and lacks feminine charms and grace. She is accused of being useless for love and a ridiculous being from whom man should keep away. The contrary is the truth. The feminist woman is an intelligent woman who wishes to disengage her economic and social standing from that of the male of the family."[26] Abella de Ramírez even argued that women should be compensated for bearing and raising children for the nation. She defended the rights of women, as well, who had children outside of marriage, and she supported equal civil status and woman suffrage—her views were considered radical by most of her contemporaries.

Catholics and other critics protested that women who ascribed to feminism risked losing their precious status in the home and turning into virtual men. Working women, paternalistic reformers further pointed out, also faced the risk of sexual abuse. Women could fall prey to men on the streets, as they traveled to and from their workplace, and to exploitative bosses in their offices and factories. Or such women might be enticed into the immoral labor of prostitution. Factory work, domestic service, and other public labor were seen as part of a continuum, with deviant sex work at the extreme. Despite feminism's moderate character, legal scholar Osvaldo Magnasco called it "foreign" and a "truly eruptive fever." He urged women to "return to the home." Manuel Carlés, another legal scholar, wrote that "women are traditional in our country, in a distinct manner than in other countries. She submits to the will of her father, obeys her husband, [and is] the grace of the home. . . . She does not freely exercise her rights, nor does she feel she needs to . . . as her legal representative [a male family member] is always devoted to defending her."[27] Most intellectuals, scientists, and politicians avoided the controversial questions of suffrage and equal political rights. Except on the matter of church power, traditionalists and reformers alike shared the same broad goal: to bolster the family as the basis of a strong society.

Regulating the Pathological Prostitute

Prostitutes, considered by state and municipal officials as morally lapsed women, and hence as beyond the normative female expectations and protections, were medically controlled and confined in an authoritarian manner. Buenos Aires of the late nineteenth century was world-famous as a city of prostitution and the "white slave trade," even though Mexico City and other cities had a higher proportion of prostitutes. In 1869 the census estimated there were 3,600 prostitutes in Buenos Aires, about 5 percent of its female population. In 1880 the hygienist Emilio Coni estimated that there were, in addition, 3,000 clandestine prostitutes. Between 1889 and 1893, nearly 4,000 new women registered with municipal authorities as prostitutes, and by 1901 the recorded number rose to over 6,400. The actual number was most likely higher. By 1903, Coni's estimate of clandestine prostitutes had risen to 10,000, in 1915 to 18,500.[28]

At the turn of the century, with immigrant men vastly outnumbering both foreign and native-born women, the demand for paid sexual services was high. Many women who could not find other jobs were forced into prostitution. Argentine women from the countryside began to migrate to the capital city and other urban centers for the work opportunities opened up by immigration, including entertainment and sex trade. Many were hired into exploitative labor as domestics or textile workers. Some foreign women, mostly Polish, Russian, German, Italian, and French women who had been lured to Argentina with the promise of "respectable" jobs or even marriage, found themselves virtual prisoners of ruthless pimps (mistakenly believed to be mostly Jewish). Many husbands forced their wives to provide sex to other men for money, and some parents sold their children for sex as well.[29]

Public houses or brothels assumed a status halfway between the public and private spheres. In 1876 there were 71 legally registered bordellos in Buenos Aires, about 250 by 1914; some were luxurious, with ample private rooms, elegant parlors, and sumptuous furnishings, whereas others were little more than a tiny room in a crowded tenement. A licensed bordello might have as few as one or two working women, or up to sixteen, but one official noted the widespread practice of underregistration and illegal boarding. In exchange for bordello owners' payment of a municipal tax of 15,000 pesos per year in 1878, the city provided medical services to the licensed sex workers and inspected the houses.[30]

Prostitution was legal in Argentina between 1875 and 1934, but many opposed it, mainly feminists, socialists, and some religious groups. The Catholic Church, however, saw legalized prostitution as a safety valve for male sexuality and did not contribute to the efforts to eliminate it. The most sustained critique emanated from the fledgling feminist movement; however, despite international pressure against the white slave trade, for many years the local campaign to recriminalize prostitution was frustrated, and opponents had little legal recourse against the trade. As in many European cities, prostitution was legalized in Buenos Aires primarily to protect its clients, not the prostitutes themselves. Governments hoped to control the spread of venereal diseases, especially syphilis and gonorrhea, which were morally and physically stigmatized and sometimes fatal, and to control the movements of prostitutes and protect neighborhoods, churches, and public buildings from contamination. Legal bordellos and an exclusionary zone were created to prevent women from soliciting in other public places, such as the downtown business district, cafés, and dance halls, and to stop women from approaching men in doorways and on the streets. After legalization, prostitutes were essentially trapped in the bordellos, where the majority earned paltry wages. They needed permission to leave the houses, and they were granted automatic permission to leave only if they were married. Women working in the bordellos sometimes kept their children with them, legal or not.

Physicians, hygienists, and city planners, following current European theory, considered prostitution the most common form of quasi-criminal female behavior.[31] Echoing the views of the influential Italian criminologist Cesare Lombroso, Antonio Dellepiane, a law professor at the University of Buenos Aires and founding member of the pioneering journal *Criminalogía Moderna*, labeled prostitutes degenerate women who lacked the ability to commit "real crimes." Prostitutes, he said, shared with male degenerates a number of traits, such as laziness and avarice.[32] The social pathologists may have disagreed about the prostitutes' immorality, and whether the practice was linked to criminal tendencies, but all were concerned with venereal disease as a public health issue. Their remedy to prevent it required confinement of the female prostitute. Enrique Feinmann, well-known hygienist and pediatrician, laid out prescriptions for prostitutes in his 1913 *Policía social* (Social policing). Illegal trade in women, he said, a "plague that . . . to the Argentines' shame has located itself in our country, and above all in the federal capital," must be tightly controlled by the nation's criminal code with jail sentences of up to ten years. Feinmann called for more officers to deport pimps, to carry out inspections at the port to identify vulnerable women who may be tricked by pimps, and to crack down on clandestine prostitution.[33]

Some criminologists went beyond criminalization of pimping and tighter regulation of bordellos to call for the protection of prostitutes through progressive social reforms like women's education, workplace improvements, and welfare measures. Even Adolfo Batíz, the former police officer and author of a picaresque study of Buenos Aires life at the end of the century, favored social legislation to improve the lives of prostitutes. He charted the sources, types, and "means of sanitizing" prostitution, and for remedies, he called on branch government, legal reform, charity, and even a "Marxist regimen," by which he meant social reform.[34]

At the 1905 South American police conference in Buenos Aires, police chief José Gregorio Rossi declared that prostitution had "produced contamination," with foreign and growing numbers of native-born women ensnared by pimps, "whose only means of living are the resources obtained through prostitution, that which they acquire by seducing, tricking, and corrupting young working women and in some cases their family members." This "grave evil" undermined the moral order of society: "it degrades men by atrophying their sentiments of dignity and morality," leading them into crime and "bringing disgrace into many homes." As a remedy Rossi urged "severe repression" of prostitution.[35] Enrique Prins, secretary of public assistance, argued that legalization itself would not reduce "perversion" among women who engaged in prostitution but did not do so out of necessity. Still, the number of prostitutes should be controlled, though "medical and prophylactic actions by the authorities" should avoid "tyrannical demands" like forced hospitalization. Prins preferred a humane and enlightened approach that encouraged education and the "moral elevation" of prostitutes; just how they were to be "elevated" he neglected to say.[36]

Yet most criminologists, like the police and the Catholic Church, believed that prostitution should remain legal in order to allow the state to monitor and contain sexually transmitted diseases. Physicians thus assumed a central role in the control of prostitutes' movements. In 1908 prostitutes were corralled into delineated red-light districts, mainly to keep them away from the downtown area, the port, and the railroad station, with weekly medical examinations required at the network of dispensaries and at a hospital for the treatment of prostitutes with syphilis (see Map 2). The prostitute patients were forbidden to speak unless in answer to a direct question from the medical staff. Women with venereal disease were to be hospitalized, by law, until cured, though this ordinance was rarely enforced since successful treatment of syphilis took two years, and there was as yet no known cure for gonorrhea.[37]

Criminologists tended to see women as victims of true criminals, the pimps. To Feinmann, pimps were "sexual parasites," or "Apaches," as they were sometimes called: "This 'Apachism' unites these beings, both men and women, who live to exploit the immoral traffic in women. . . . They organize themselves in capitalist companies, in secret agencies and financial societies." The "sinister" ones lured youths "with false promises of easy work, a better life or marriage" and then confined them "through guile or force in houses reserved for the purpose." Thus, wrote Feinmann, "the woman is a slave in every inhuman and agonizing sense of the word." The prostitute's status "stripped her of her natural role as the 'good fairy' " of the home.[38]

Still, prostitutes and domestic servants, as well, were considered more likely than other women to commit the worst offenses against motherhood known to society: abortion and infanticide. These offenses were widely practiced in the absence of available contraception and as a result of high rates of rape and corruption of minors and young female servants. Infanticide itself was often understood as a way of preserving the "honor" of an unmarried woman but also, increasingly, as a sign of psychic disturbance, the "quintessential crime against motherhood."[39] As psychiatrist José Ingenieros reported in a case study from the early 1900s, a young prostitute who killed her newborn was a "morally insane child-killer," and her lack of remorse led doctors to repeat that "in this patient there are no anomalies of the sexual emotion . . . but rather in the very sentiments of being a woman."[40] Women were rarely convicted of these two crimes, however, and between 1871 and 1905, only twenty, the majority of whom were domestic servants, were convicted of infanticide. Most of the twenty received sentences of three to six years, a much lighter sentence than that for homicide. One solution offered in 1888, in an attempt to reduce immoral behavior among domestic servants, was a passbook, a *libreta*, issued by the city government to every maid. Each girl or woman was required to register with the city and carry this book with its certificate, signed by her employer, attesting to her good conduct and work history.[41]

Those few infanticide and abortion trials that did occur were colored by age-old ideas about women's agency. Police entered the accused women's homes and gathered bloody sheets and clothes, sharp instruments, and toilet fixtures. Medical inspectors were called to examine the women's reproductive organs for signs of recent childbirth. The bodies of the dead infants (if found) were autopsied and dissected. As in other criminal cases, forensic experts testified at trial. But ultimately, at these trials women were held to a different moral standard from that applied to male murder defendants. A woman could use the defense of her womanly honor to explain the illegal death of her infant. The act of hiding a pregnancy and discarding an illegitimate newborn to protect her family's or her own reputation was the most approved choice.

Medical expert witnesses played an ever more central role in the trials of women accused of infanticide. In 1912 a twenty-four-year-old servant "girl," Teofila Acosta, felt back pains and retreated to the toilet to relieve herself. According to her own version of events, much to her surprise she gave birth to a stillborn infant. After the eventual discovery of the fetus in the WC, Acosta was brought to trial. Experts were summoned to determine a number of issues: had she given birth or induced an abortion? Was the child

born alive or dead? Was the mother ultimately responsible? After an autopsy, the physicians related that, typical of these cases, they could not
answer the judge's questions because of the late discovery and advanced
state of decomposition of the body. Based on the lack of conclusive physical
evidence and on the testimony of her employer, Señor Caminos, of Acosta's
"irreproachable conduct and service" in his household, she was acquitted.[42]
Despite Catholic Argentina's claim to cherish children, and its desire to
rid society of crime, outdated ideas about women's roles and morality
persisted.[43]

The Civilizing Influence of Mothers and the "Improvement of the Species"

Although religious and scientific figures alike considered women's natural
state to be marriage and motherhood, children, the product of those institutions, were even more important than mothers. Criminologist and psychiatrist José Ingenieros's book *Tratado del amor* (Treatise on love) called the
propagation of the species humanity's ultimate evolutionary goal.[44] The
primacy of motherhood was a nearly universal ideal, and contraceptives,
available in other countries, were illegal in Argentina, the very topic scan-
dalous. The nurturing of children was seen as women's most natural calling
and noble labor.

Women who rejected motherhood were regarded as barbaric and immoral; civilized women understood their maternal role well. Women were
said to have a "civilizing and beneficial" effect as mothers, with a responsibility to fulfill their natural duties. One woman who refused sexual relations with her husband and repeatedly proclaimed that she wanted no children was diagnosed with "maternal phobia." Her husband reported that she
found "the noble maternal desire" repugnant and that she threatened to
leave him because of his "legitimate desire to be a father"; he concluded
that "neither reason, nor love, nor all my demands, can do anything against
her."[45]

In 1903 a case was brought to court by the family of Señora "S.G.," a
young widow afflicted by mental illness, characterized by a lack of "moral
sense," and abuse and abandonment of her family. "To many questions of a
moral order that we set forth, she answered in a precise manner," said the
medical assessment. "She understood her mission as mother, her duties in
the education of her children, also in the running of the household . . . but
she ignored all that and could not appreciate their worth."[46] Two medical

experts, Carlos Benitez of the National Department of Hygiene and Juan
Acuña of the court, diagnosed severe hysteria and fantasies of persecution.
They pointed out, "We know very well the dangers to which these mentally
ill patients are exposed, and the harm that they can cause, quickly convert-
ing the persecuted into persecutors, and bringing them to these conditions,
even criminality." They provided a lengthy family history, a discussion of
S.G.'s inherited tendencies to madness, her menstrual history, and her biotypology, as well as her coloring and appearance ("green eyes and abundant
black hair. Her face and skull do not present stigmas nor asymmetries that
differentiate her from the normal anthropological type, and although her
physique has suffered much, one can still appreciate in her features the
delicate lines of her former beauty.").[47]

Officials noted a relationship between "women's fecundity and the dura-
tion of their families. . . . The contest of the crib is, in effect, one of the most
important factors of the 'struggle for life' of nations and races." The first
step in determining whether a nation's fertility was healthy was to count the
number of women, especially married ones, and the number of children,
and from what ethnic source they came. Argentina, in the view of census
officials, had special characteristics that needed to be taken into consideration: "Dealing with a nation like Argentina, with its fusion of all races of the
earth that converge there, attracted by the perspectives of well-being and
fortune afforded to them, one should study the fertility of women in each of
the foreign nations that have moved or formed their homes here."[48]

Argentina had one of the highest birthrates in the world in 1914, reaching
nearly 37 births per 1,000 inhabitants, compared to 17.7 births in France,
23.8 in Britain, and 30.3 in Italy. Among the 606,174 women surveyed for the
third census, native-born Argentines had the highest fertility rates, and
Italian women had the largest families. The average number of women per
children was about 4.3 in 1914, with Spaniards having the lowest number at
3.7, Italians the highest at 4.9. Census officials were nonetheless concerned
that birthrates were declining. Charting the number of married women
with no children, they reported that infertility was rising. In 1904, 11 percent of women had no children, but by 1914 the percentage had reached
14.8 percent.[49]

In 1904 President Julio A. Roca drafted Juan Bialet Massé, a Spanish-born
physician and educator, to survey the conditions of the working class in the
interior of the country and in the capital city and compare them to those of
other "civilized" countries. Bialet Massé's findings ranged from studies of
living conditions in the city slums and in rural shacks, to the work environ-

ment for men, women, and children in factories and on farms, to education for the children of the poor. Bialet Massé considered the home the source of the high rates of tuberculosis among the city's poor and blamed the "misery of work of women and children" and "an insufficient wage." Women's practice of washing the clothing of the ill along with that of uninfected family members would also spread the disease.[50]

Even worse, according to Bialet Massé, women who left home to work created a "morally and materially unhealthy atmosphere" that jeopardized their own and the family's health. Work outside the home could "disturb or ruin" the "high function of maternity." "Woman's mission," Bialet Massé wrote, should center on "the perpetuation and improvement of the species." Professional women were a "third sex" who "remain without men. . . . [T]hey have renounced marriage, and in their delirium arrive at castration." The professional woman, in numbers greater than ever in Europe and the United States, "happily does not exist yet among us, except for a few individuals." Yet, recognizing that combining home with work was probably an inevitable trend of modern times, Bialet Massé recommended a pragmatic mixture of social reform and punitive measures. He supported protective legislation that would limit working hours and guarantee equal wages and safe working conditions for women and children and that, for both men and women of the laboring classes, would improve nutrition, work safety, and educational and cultural opportunities (the "civilizing effect of the theater," for example, could replace harmful activities). Bialet Massé, reflecting commonly held views, recommended that women already in the workshops and factories have shorter work days and safer conditions, but he preferred that women not work outside the home at all. Diseases such as tuberculosis were widely understood as caused by overwork; women were considered more susceptible to such illnesses than men. Criminology professor Cornelio Moyano Gacitúa wrote in 1913 that a woman's "sexual condition places her in the home. . . . [H]er delinquency progresses with her advance to employment, to the factories, to the streets." Women were more vulnerable to moral corruption than men and could be unwittingly sucked into a life of crime or prostitution.[51]

Scientists and medicolegal experts firmly placed women in this home so they could manage and nurture their families. One local judge declared, "The disorganization of the family is a general modifying cause of the moral state of the child."[52] In 1913 the *Archivos* published an unsigned review of a German book that linked child abandonment with inborn adult criminality and referred to the "explanation of causes of moral degeneration in children that the author relates to psychological and biological conditions of their forebears."[53] A Russian study on "children of criminal and noncriminal parents," reprinted in 1899 in the *Semana Médica*, divided 116 girls and one boy into groups of noncriminal children of criminals, noncriminal children of noncriminal parents, and criminal children of criminal parents. Concentrating on the sexual development of the girls, the author examined the onset of menstruation, sexual activity, and masturbation and concluded that the children of criminal parents were "morally less developed" but "physically more developed." Criminal girls were identified with precocious, and possibly promiscuous, sexuality. The author recommended regular and obligatory studies of this sort in all schools, "in order to determine, as quickly as possible, the moral and physical defects of children and to take the appropriate medico-pedagogical measures." He suggested segregating the "physiologically and morally weaker" children.[54]

A 1916 article in the *Revista de Criminología, Psiquiatría, y Medicina Legal* (Review of criminology, psychiatry, and legal medicine) on "women's growing impotence to nurse their children" warned that the decline in breastfeeding was contributing to the degeneration of the race. Its author, a Swiss professor of chemical physiology, attributed to the lower rates of breastfeeding "symptoms" of social pathology like the "diminution of resistance to diseases such as tuberculosis, nervous diseases, and dental cavities. Children are poorly nourished and thereby degeneration increases from generation to generation, and brings as its conclusion the endless jeremiads of the disappearance of the species." Since there seemed to be an "inherited incapacity to nurse, when a mother cannot sufficiently nurse her child during the normal period (that is, a whole year or at least nine months), the daughter equally and almost without exception does not possess the normal faculty to nurse." Because a mother's responsibility to nurse was so great, "men of good health, desiring of healthy descendants," should marry only a woman who had herself been breastfed as an infant.[55]

The medical focus on the family served a vision of children as the future citizens of the nation and as an advancing race, and it was a reaction to an apparent epidemic of crime and other forms of antisocial behavior among youths. Miguel Lancelotti, noticing an alarming rise in juvenile delinquency, blamed both society and family disintegration for it.[56] Eusebio Gómez, in *Mala vida en Buenos Aires*, presented photographs of juveniles to illustrate the various "types" of urban thief.[57] Because youths were impressionable and vulnerable to environmental influences, often lacked a "proper" family setting, and presented special dangers of reincidence, José

Luis Duffy, director of the boys' reformatory in Buenos Aires, proposed that the state take over their education. Duffy sought a tightening of the law to prevent juveniles who were freed from jail from falling back into crime and suggested a police investigation of the minors' parents or guardians. "If they prove that the family is honest, the minor will be sent back to them"; if not, the minor would be sent to the reformatory in Marcos Paz and kept under the tutelage of the Defensoría de Menores.[58]

Prison physician Eleodoro Giménez, after examining scores of medico-legal profiles of delinquent youths, concluded that the principal factor in juvenile crime was an "educational lack in the family, the little importance that parents give to the education of their children." Questions of class came to the fore. Giménez mentioned abandonment of children, poverty, and "hereditary defects" and cited the lack of family unity and cohesion, of good habits and faithful spirit in most "working families," traits not "exclusive to our nationality" but rampant in Argentina at the time.[59] Similar studies appeared with greater frequency over the next decades. A 1920 article focused on poverty as "an enemy of child health" but said a child of affluence could be "vain, class conscious," and nepotistic. This article held out the "child of the middle class" as "the best situated to physiologically form the mentality of the 'normal man.'" Active state intervention in education would assimilate and homogenize children of all classes and teach them proper social conduct. The author argued that "education can correct the influence of the family environment in all cases; therefore it is desirable that children's education should be controlled by the State, whose functionaries, shaped by science for life, are capable of modifying the dark shades that contemporary reality offers to children of various means."[60]

In 1912 the *Archivos* published an article by criminologist H. Leale on women's contribution to raising moral and law-abiding children. "Those who affirm that the woman is less criminal than the man forget that she forms the soul of the children more than the father," the author wrote. "There is a very well known fact: the mother transmits and feeds her vices and her virtues to her son. The boy, the citizen, the man, are all formed by the mother. To say that the man is a criminal is to say that the mother was one. . . . All this leads us to the conclusion that 'the woman cannot be less criminal than the man.'"[61] Victor Pesenti's thesis for his law degree at the University of Buenos Aires proclaimed that "the child derives unawares his habits and ideas in great part from the models and stimuli that he receives in the home." In Pesenti's view, "modern life" presented an "evil influence" on the home, causing sexual perversions and even abortion and suicide.[62] Mothers could prevent criminality in their children. As Leale wrote in 1912, "Is the woman less of a hero? No! Her heroism is hidden in the home . . . . [To] be a mother is the greatest heroism. Not all men are heroes, but all mothers are. This serves to clarify the question of criminality of the sexes."[63]

Physicians knew that both men and women could influence heredity and the health of their offspring, but they considered women accountable for the propagation of the species because of their unique biological capabilities and their social roles as nurturers. Most important were the care, education, and model they provided to future citizens of the nation. In addition to moral homes, Pesenti prescribed better and longer education for "the cultivation of the spirit." Women, as both mothers and the "natural" teachers of children, were to guarantee their children's attendance in school. School was a "means of prevention against delinquency" and a way to "cement moral ideas." At the most basic level, social scientists, psychiatrists, and medicolegal experts placed the moral responsibility for reproduction of culture and morality on women. A 1903 European article on women's roles, reprinted in the *Archivos*, suggested that women have fewer children (though it did not openly endorse the use of contraceptives), so they could educate and properly raise the ones they had. Said the author, Judge Aubanel, "Fertility should not be measured in the number of children, but in the quality of them." He wrote of "a certain antagonism between physical fertility and morality, so that the more children a woman has, the less she can occupy herself with the development and education of their intelligence. Therefore it is important that the woman enlighten herself, perfect herself, and become capable of giving to her children a vigorous and useful moral education without worrying about [her] fertility, and taking into account only that the few children [she has] will be robust and intelligent."[64] Not only the number of offspring, but their degree of health and fitness were now of concern.

In the late nineteenth century, infant mortality, generally considered an indicator of a nation's health, was high in Argentina compared to most of Europe, but down from its peak of the late 1860s, when, in Buenos Aires, infant deaths had reached nearly 30 of every 100 births. By 1905 that rate had fallen to just under 10 deaths per 100 births, the result of improved sanitation, especially running water, sewers, and public health services.[65] Although women were considered naturally talented at nurturing children, scientists introduced the idea that good mothering could be produced by a scientific regimen. After 1900, hygienists, gynecologists, and pediatricians in Argentina advanced a new specialization of *puericultura*, or puericulture (literally, the science or practice of growing children). Genaro Sisto, a prominent professor of pediatrics at the University of Buenos Aires, called puericulture the "science of the moment."[66] Originating in France, puericulture was intended to instruct mothers, female health care professionals, and teachers; female medical and nursing students were encouraged to specialize in puericulture or related fields like obstetrics.

Feminists, among them some of Argentina's first women physicians, began to devote themselves to such work. Cecilia Grierson, who was a secondary school teacher as a young woman, then the first woman physician in Argentina and a feminist activist, took on the health and well-being of women and children as her primary concern. At age twenty-three, she enrolled at the University of Buenos Aires medical school and graduated in 1889 with a specialization in obstetrics. While still a student, Grierson helped to establish the university's nursing school. She became its first director, and the school was named for her in 1935. Grierson taught obstetrics at the medical school and later worked in kinesiology and emergency medicine at a number of Buenos Aires hospitals. She presided over the First International Feminine Congress of Argentina, held in Buenos Aires in 1910, and contributed to efforts to reform the civil code in favor of women's rights. Elvira Rawson de Dellepiane, a physician colleague, also at the congress in 1910, urged better nutrition and medical care for poor children. Other participants suggested more widespread efforts to train women in their early childbearing years in the latest scientific approaches to child care.[67]

Feminist and socialist reformers championed issues like prenatal leave for working women and mandatory breaks for nursing mothers, and they proposed protectionist legislation to reduce infant mortality and debilitating disease and to ensure equal wages for women. They also sought improved conditions in factories. One socialist document of 1894 called exploitative female and child labor in factories "the source of many ills in the family."[68] Buenos Aires factories, by all accounts, were poorly lit and ventilated, and rarely clean, with women and children workers forced to sit at repetitive work for hours on end and denied bathroom breaks. Some industries, such as match making, used noxious chemicals. Gabriela Laperrière de Coni, an active social reformer and wife of the hygienist Emilio Coni, warned that the substandard environment of Argentina's factories would destroy the reproductive health of women and the health of their future children.[69]

The leading pioneer of the new puericulture was Enrique Feinmann, a Buenos Aires pediatrician who focused on the prevention of venereal disease and who had a related interest in crime and delinquency. Feinmann insisted that the nurturing of healthy children was one of the best ways to defend national progress against the danger posed by abandoned and delinquent youths. Following French innovations in this field, and from his faculty positions at the University of Buenos Aires and at the city's elite boys' high school, the Colegio Nacional, Feinmann publicized and distributed his prescription for the "social defense of the child." He recommended "social prophylaxis" of children to "resolve in our time this problem of civilization facing the country." He urged an expansion of reformatories and special institutions for children, the modernization of education to inculcate physical and spiritual health, and the passage of vigorous legislation to protect "maltreated, abandoned, and mistreated juveniles," a "world of tiny social members."[70]

As the idea of puericulture caught on in Argentine medical circles, Feinmann, along with physicians Grierson, Rawson de Dellepiane, Gregorio Aráoz Alfaro, and others, began to issue chemical analyses of breast milk and formula and instructions for proper swaddling and dressing of infants. They intended puericulture to be incorporated into girls' curriculum in public schools. In 1913 the First National Congress of the Child was held in Buenos Aires, one of the first scientific gatherings in Argentina where women professionals such as Cecilia Grierson could participate. The state welfare apparatus, or Asistencia Pública, had created in 1908 a department devoted to the protection of children under two years old. The Protección de la Primera Infancia, as it was called, established and oversaw milk dispensaries, visited nursing mothers and provided them with food, and after 1919, examined wet nurses. The number of children helped by the agency expanded from 232 in 1908 to nearly 12,000 by 1920; even more numerous were the women who had contact with the Protección de la Primera Infancia, which made more than 500,000 consultations or inspections in 1916.[71]

Hygienists were intent on reducing the infant mortality rate but even more on eliminating "degeneration of the race" caused by faulty or ignorant mothering. Anxiety about depopulation in Argentina built on the idea that Argentina was an "empty" land in need of an influx of numerous, healthy inhabitants who could enhance the nation's wealth. Feinmann, with the optimism of the age, had a vision of fertility and health for Argentina. He wrote in the introduction to his 1915 textbook *La ciencia del niño* (The science of the child), "Woman will be the good fairy of the new era. Her nursery of human beings will be an immense blooming garden, and the children, instead of going to heaven as angels, will populate the earth as men, to make it better and more beautiful."[72]

Newly arrived immigrants entering the Hotel de Inmigrantes (Immigrants' Hotel) in the port of Buenos Aires, early 1900s. (Archivo General de la Nación)

Immigration as a sign of progress: "Immigration as a means of progress and culture." Drawing of the ideal immigrant in the sophisticated Buenos Aires weekly magazine *Caras y Caretas*, 1 January 1914.

La inmigración como medio de progreso y de cultura

Inmigración peligrosa

Pesquisa 1.° — Tiene razón el ministro de agricultura. Hay que seleccionar bien á los inmigrantes, como en Norte-América.

Pesquisa 2.° — ¡Ni que hablar!... Y fíjate en aquel. Vamos á averiguarle si ha estao loco y si sabe leer y escribir.

Dib. de Zavattaro.

Immigration as a sign of danger: "The immigrants must be well selected, as in North America." "Right! . . . Let's check if he is insane
and if he knows how to read and write." "Dangerous Immigration,"
a cartoon in *Caras y Caretas*, commenting on the selection of immigrants to Argentina, 12 June 1909.

Spanish immigrants in the port of Buenos Aires awaiting medical inspection in the late nineteenth century. (Reginald Lloyd, *Twentieth-Century Impressions of Argentina*, 1911)

José María Ramos Mejía, a leading psychiatrist and member of the Generation of 1880. (Archivo General de la Nación)

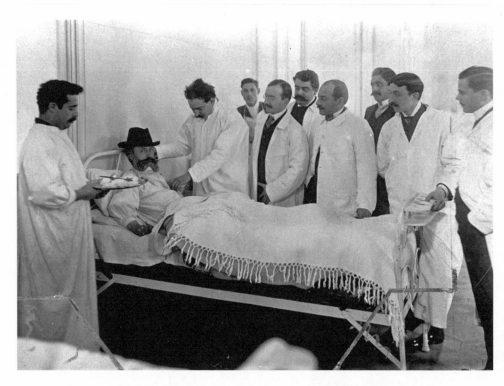

The criminologist José Ingenieros, third from left, examining a patient as colleagues look on, early 1900s. (Archivo General de la Nación)

"The manner of applying the swaddling clothes": Maternal and infant hygiene as the modern way to care for infants. (Enrique Feinmann, *La ciencia del niño*, 1915)

A children's correctional school, early 1900s. Education was seen as key for the rehabilitation of both children and adults. (Archivo General de la Nación)

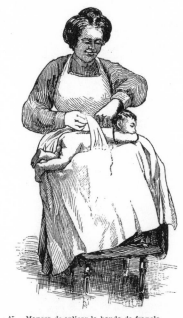

45. — Manera de aplicar la banda de franela.

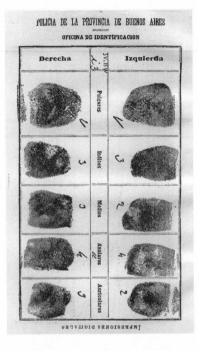

POLICIA DE LA PROVINCIA DE BUENOS AIRES

OFICINA DE IDENTIFICACION

One of the first criminal identification cards from the provincial police of Buenos Aires, 1891. (Luis Reyna Almandos, *Origen del Vuceticismo*, 1909)

Italian criminologist Enrico Ferri, center right with plaid coat lining, on a visit to Argentina's model prison, the Penitenciaría Nacional de Buenos Aires, 1907. (Eusebio Gómez, *La Penitenciaría Nacional de Buenos Aires*, 1925)

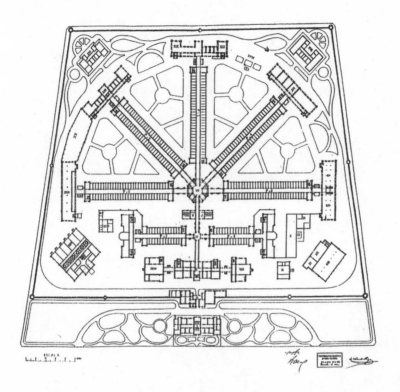

Layout of the Penitenciaría Nacional, with five pavilions organized radially, based on Jeremy Bentham's panopticon. The central watchtower is marked as VII. (Antonio Ballvé, *La Penitenciaría Nacional de Buenos Aires*, 1907)

Prisoners at sewing machines at the model Sierra Chica prison, 1928. Inmate labor, intended to foster work discipline and morality, was a key concept in the penitentiary. (Archivo General de la Nación)

A patient subdued by guards at the Hospicio del las Mercedes, a public mental institution in Buenos Aires, early 1900s. (Archivo General de la Nación)

Buenos Aires police removing a drunken woman from the street, 1902. (Archivo General de la Nación)

Buenos Aires police marching against disorder, 1923. (Archivo General de la Nación)

Lucas Ayarragaray, national legislator and
author of xenophobic measures in the early
1900s. (William Belmont Parker, *Argentines of
To-Day*, 1911)

SIX

Men on the Street

A Threat to "Our Industrial and Social Organization"

> The law should intervene energetically against the parasites [i.e., vagrants and beggars] that systematically refuse to obey the natural law of work, especially since their conduct, generally depraved, constitutes a true focal point of moral infection for the people with whom they are in contact.
>
> —BELISARIO J. MONTERO

In February 1911 Buenos Aires chief of police Luis Dellepiane announced a program to eliminate alcoholism among men, a cause of "so many great disturbances of all types." The new ordinance would require all bars to close on Sundays. Dellepiane noted with disdain that many working men spent their entire free time getting drunk. The consequences were brawls, violence, visual and noise pollution of the cityscape, and neglect of the men's families. The official dispatch pointed out that "the moral goals of the law of Sunday rest will be stripped of their virtue" if workers headed for the bars on their day off. By limiting the availability of alcoholic drinks, "the worker would gain in health and the elevation of his social and intellectual level," and not "lose it in the tavern, an incubator of vice." The daily paper *La Prensa*, reporting on the ordinance, quoted Dellepiane that "prescribing this law" was crucial to "repressing alcoholism, the principal cause of crime."[1]

In cafés, bars, and popular watering holes called *pulperías*, working men congregated to drink, smoke, play cards, sing, dance, and fight. Juan Bialet Massé, a Spanish-born physician and educator, in a 1904 report on the conditions of the working class, identified bars as the gathering place of "those who protest, who make meetings and loudly shout to strike [at] liberty." The owners of taverns "enrich themselves exploiting and fomenting the vices of the poor."[2] Many bars and dance halls were poorly disguised houses of prostitution, and to the city's social pathologists, prostitutes were intermediaries to "true" forms of crime, whose seedy locales attracted men of real vices and degenerate tendencies. Male unruliness would erupt from the red-light district to cause disorder and disruption in the city at large. Even tango dancing, a popular pastime in bars, cafés, and the streets, was considered unsightly by elite observers. A mix of African, Spanish, Italian, and German music and dance styles, the tango had emerged from the ghettos of Buenos Aires in the late nineteenth century; although originally a dance for two men, by the 1880s it had become associated with prostitution, to the extent that this period is called the "bordello era" of tango. Only after 1910, when tango became a fad in France, was it accepted by the upper classes of Argentina.[3]

The home may have been seen as the root of all behavior, good and bad, but the street was a dangerous place and the location of most crimes. Offenses in the streets, in cafés, and bars totaled about 60 percent of all crimes between 1885 and 1891. Arrests for disorderly conduct and unauthorized use of firearms (and excluding cases of public drunkenness) averaged more than 12,000 per year between 1896 and 1905. Arrests for drunkenness alone numbered, on average, 17,000 per year between 1900 and 1910.[4] Street disturbances challenged the social order and threatened core economic relationships. Accidents were on the rise, as was violence. As increases in police budgets and numbers of officers earned state and municipal support, the public sphere became a site of accelerated regulation, policing, registration, and segregation. Liberal reformers sought order and voiced the opinion that poor and crowded neighborhoods should be regulated and protected, slums sanitized and controlled. Prostitutes were a necessary evil but must be contained and regulated, lest they draw men into the criminal underworld. Medical inspections, examination, and surveillance of public houses should be conducted. Beggars and vagrants should be eliminated, or at least redirected. Crowds should be dispersed with curfews at times of unrest, and access to alcohol should be limited. Labor should be regulated and welfare provided in order to reduce class conflict. Should these precautions fail, police action would take over.

At the heart of such concerns about the urban streets were idle or unproductive men, some of whom even sought to destroy the state itself. Women, seen as fundamentally flawed but incapable of causing great harm to the larger society, were kept at home, or sent home, or shuttled by physicians and judges to the church and Beneficent Society for moral treatment and discipline. Although government officials were concerned with women's pathology, they were much more concerned with the male sphere, the street, which was occupied, as they saw it, by scores of degenerates of both sexes.

Embedded in state officials' plans for controlling and disciplining public behavior was the hope of a restoration of national character that was intimately tied up with Argentine femininity and manhood, both under siege from multiple social forces and individual weaknesses. Men's inability to avoid toxic ideas, substances, and behavior presented three outcomes: the production of unhealthy children, the wasting of labor power and squandering of the nation's capital, and the outright threat to the state presented by anarchism. Social pathologists, aware that children were the product of fathers, not mothers alone, held up expectations of male behavior as proper providers for the family. In accordance with popular views, they saw men's role in the family as providing for their dependents and passing on healthy inheritance. Legitimate paternity and economic support were basic. So were fathers' roles as models of the work ethic for their sons. Toxic behavior, such as excessive drinking, gambling, or abandonment, which could lead to children on the streets and children's absence from school or work, were taken as factors in juvenile crime.

Productive labor was key to the "Argentine way of life." An 1899 article in the *Revista de Policía* (Police review) said, "Work is the only means of regeneration of the vagrant."[5] Diplomat Belisario J. Montero, a regular contributor to the *Archivos de Psiquiatría, Criminología, y Ciencias Afines* (Archives of psychiatry, criminology, and related sciences), wrote of "social parasites" that "it is necessary to employ severe and rational means, and . . . create poor houses where, applying a logical disciplinary system, they are obliged to work."[6] Officials sought to control the moral contagion that diminished productivity, disrupted the social peace, and could spill out of working-class neighborhoods. Government officials were not only concerned that crime was likely to retard national progress but also, and even more terrifying, that it could provoke disruption of the social order. Recent immigrants now wanted part of the national wealth and, even worse, had massed into a growing and vocal labor and left-wing movement. Protection of social order became the state's and the elite's top priority. Short of a radical investment in social improvement, the state opted to invest in repressive policing and regulation, exploiting the latest techniques to control men's movements and behavior. The church, which surveilled the daily behavior of private lives, especially of women, was less directly involved in the male world of the public sphere. Science assumed the role of devising means to support the state's control of men's public lives.

Social Parasites Who "Refuse to Obey the Natural Law of Work"

To social pathologists alcohol was the singular vice of the working man. The Italian journalist Pietro Gori, who was living in Argentina, wrote in 1902 that alcoholism was a particular danger in Latin America because liquor was abundant.[7] Another article gave alcoholism the "first place as a primordial agent of degeneration" and depicted a direct correlation in "civilized countries" between alcohol consumption, madness, and crime.[8] A sanitarium director, Fermín Rodríguez, linked alcoholism to men of the working class, quoting the French author Georges Vonjean: "In sum, this fatal passion degrades the individual, ruins the family, and destroys childhood. From it comes the diminishing of the population, of work, of wealth; in one word, of the very power of the nation!" Psychiatrist Joaquín Durquet addressed Argentina's specific situation: "In a young nation like our own, which generously offers land to working men . . . it is a holy task to oppose the causes of premature decadence, fighting the imminent threat of advancing alcoholism, by attempting to decrease its consumption and increase production."[9]

The effects of alcoholism on death rates and the nation's economy were not the only concern. Physicians also feared the long-term effects of alcoholism on the "germ plasm" of the nation's offspring. Durquet believed that a father's alcoholism would lead to madness in his daughters, making them "hysterical, neuropathic, epileptic, weak, immoral, imbeciles, or idiots."[10] A 1913 study by Swiss psychiatrist G. von Bunge, published in the *Archivos*, on the "sources of degeneration" warned of the social costs of alcoholism: "one man in ten dies of *alcoholism*, leaving behind deformed and degenerate children."[11] Alcoholism, as the worst male vice, was seen as affecting family life, children's heredity, and the stability of the larger neighborhood environment. Criminologist Francisco Netri saw links to adultery and the disappearance of the family: "Alcoholism, which almost can be considered a painful effect of our industrial and social organization, predisposes irresponsibility, lack of will, and leads to violence. It is known that adulterers are numerous . . . and if you add to this that the children and women are obliged to exhausting labor in the factories to meet the necessities of life, one easily understands why the concept of the family is nearly disappearing."[12] Criminologist Héctor A. Taborda, associate of the new Criminology Institute, called alcoholism "one of the plagues that currently eats away at

humanity and in particular at the working class of the most civilized countries."[13] The 1913 *Archivos* study by von Bunge identified alcoholism as the
primary cause of degeneration that was transmitted to offspring by fathers.
In good families, "in which mother and daughter both possess the normal
ability to lactate, it is exceptional to find a drunk or drinking father." More
than 70 percent of degenerates in his study had a father who was a heavy
drinker, and 40 percent had fathers who were "confirmed drunks."[14]

Writing in their capacity as physicians in the Buenos Aires city police's
mental observation service, psychiatrists José Ingenieros and Juan Córdoba
had concluded that alcohol exacerbated inborn tendencies to crime and
mental illness. "F.N.," a mentally ill alcoholic, was "dangerous . . . which
makes him inadaptable to the domestic and social environment in which he
lives."[15] Of another alcoholic, Ingenieros and his colleague Lucas Ayarragaray reported a diminished moral sense and intelligence and "well-defined
degenerative traits."[16] A Peruvian medicolegist, Leonidas Avendaño, saw
alcoholism as a cause of epilepsy in subsequent generations. "The struggle
against the devastation of alcoholism," he said, "seriously preoccupies governments and intellectuals in nearly all nations."[17] Chilean doctor Joaquín
Castro Soffia blamed parents for passing on both a toxic environment and
defective physiology to their children: "Alcoholism also plays a principal
role in heredity, since the children of alcoholic mothers and fathers are
often epileptics, idiots, or degenerates, and are predisposed to alcoholism
and madness."[18] A review of a Uruguayan book on alcoholism commended
the author for "putting in relief the alarming increment of alcoholism in all
civilized countries, making special reference to Latin American nations,"
and praised his call for social measures to eradicate the disease.[19]

In his book *Policía social* (Social policing), Enrique Feinmann, a pediatrician who pioneered the field of maternal and infant hygiene in Argentina,
labeled alcoholism a "historical, moral, and biological vice." He found in
as many as 60 percent of Buenos Aires city hospital patients a "plague [that
is] growing in alarming proportions . . . a grave social danger." On par
with pathogens like tuberculosis, alcoholism threatened the "destruction of
modern nations. . . . It not only produces serious damage and multiple
disorders in the individual, but also affects the species [in terms of] the
birthrate and descendants. In effect, the alcoholic diminishes the birthrate
and causes degeneration of the species." The last stage of chronic alcoholism, Feinmann warned, would result in "imbecility, mental retardation, and
sterility with the extinction of the family." Alcoholics could breed depraved
criminals, like the murderer Cayetano Santos Godino, "son and brother of

alcoholics, himself an alcoholic at sixteen years of age," and their female children were in danger of becoming prostitutes because "their weak consciences result in their suffering a lack of moral sense."[20]

Despite widespread alcoholism, the only treatments available, Feinmann lamented, were "symptomatic and deficient." He did suggest the creation of anti-alcohol leagues and legislative measures such as propaganda and education aimed at children in the public schools. He recommended state control of public drinking, including raising the legal age for purchase of alcoholic beverages and criminalization of serving visibly drunk customers. Police control was also necessary to clean up the "immoral spectacle" that drunkards displayed to the public. Special attention needed to be paid to workers, as among them "the danger is deepest, attracts most often, and is most harmful." The "popular classes" were particularly in need of instruction in hygiene and morals, for the "spiritual and physiological regeneration of the worker."[21]

There was a range of opinions on how to deal with alcoholic men. To diminish the high percentage of poor men with alcoholism, some experts, such as von Bunge, prescribed state regulation, even prohibition, since alcoholics were incapable of drinking moderately on their own. If alcohol was excluded as the fundamental cause of disease, "it will be easy to discover other causes of illness and confront them at the same time with appropriate remedies." Liberal social reformers' and charities' approaches, in von Bunge's view, only perpetuated the problem. Socialists wanted "to organize the state of tomorrow, but do nothing to remove man from degeneration and decay."[22] In his influential *Nuestra América* (Our America), Argentine sociologist Carlos Bunge argued that a deficient work ethic was the principal problem facing Latin Americans in general, and Argentines in particular. His recommendation was simple. "First of all, let us work! First of all, above all, *for* all, let us work! Do not object to me saying work to a lazy person is like saying to someone who is ill that he should be healthy." A new work ethic would make indolent Argentines more like their industrious, northern Anglo-Saxon cousins. In Spanish American society, "therapeutics of creole politics" would create "culture through work" and counter the "psychological and sociological state produced by creole inertia."[23]

Alcoholism in men could lead not just to disease and death but to harmful conditions for society. A high proportion of alcoholics "do not want to work: beggars, addicts, bums, and vagrants." Belisario Montero urged the law to "intervene energetically against the parasites that systematically refuse to obey the natural law of work, especially since their conduct, generally depraved, constitutes a true focal point of moral infection for the people with whom they are in contact."[24] According to an Italian author in the *Archivos*, "the vagabond, in general, because of the neurasthenia that invades him and makes him inadaptable for civilized life . . . fatally reproduces in all times and in all nations, two primitive characteristics of man that can still be found in many savage tribes: inertia, with the incapacity for regular work . . . and the tendency to wander and lose his way." This author identified numerous varieties of vagabonds: "social refuse of all types, poor and ill rootless ones, rejects, *declassés*, elderly troublemakers, idlers, unfit, alcoholics, impudents, obstinate workers . . . and finally those depressed by poverty," as well as delusional maniacs and those in a psychological "vagabond crisis." He provided psychiatric classifications like the "migratory degenerate" to identify and handle "anomalous" and "socially segregated" individuals.[25]

Streets in poor neighborhoods and the downtown area were filled with vagrants and beggars and other so-called social parasites. Boys, sometimes as young as three years old, ran in packs. Alberto B. Martínez, the city's director of municipal statistics, described "ragged and noisy children" who "sell papers, clean boots, run errands, etc. . . . only because a pittance can be earned by them." The children in the streets were one result of sudden immigration and urbanization. Shantytowns had sprung up on the outskirts of the dump, near the port, and even in the suburbs—huts of tin with no sewers, electricity, or running water. Martínez pointed out that the government preferred newcomers to settle in the interior rather than clog the city, but "it can do no more than advise and cooperate. It cannot use force."[26] Francisco de Veyga, the nation's leading medicolegist, told the Argentine Psychology Society that "the thief, inept since childhood for social life and refracting all culture and discipline, begins his criminal career as a vagabond minor, receives professional sanctification in prison, living later between jail and street for the rest of his life, without modifying his status nor his aptitudes." At the root of this life were degenerative heredity and a deficient intelligence, compounded by illiteracy. Psychological examinations of such individuals revealed one constant characteristic, "the absolute incapacity for reflexive work."[27]

In the political and legislative sphere, social welfare and reform proposals were supported primarily by the Socialist Party and Social Catholics, the critical, progressive, and political wing of the Catholic Church. However, few reforms to improve the wages and services to the poor were implemented. A much-discussed labor reform bill introduced by President Roca in 1904 to mandate a forty-eight-hour work week and additional pro-

authority." Nunes explained that the Province of Buenos Aires was "composed of men from all parts of the globe, who belong to various nationalities and distinct races, with ignorant life habits, native tendencies, peculiarities, [and] characteristics . . . whose forms are consequently unknown to the majority of agents. This difficulty, in most cases, can cheat success in important investigations."[35]

There had been a series of social upheavals, frequently violent, early in the new century that had amounted essentially to class war in Argentina's major cities. For the vast majority of Buenos Aires's population, living conditions were poor, salaries were low, and workers in nearly all professions clamored constantly for higher wages. Worker militancy after 1891 led to repressive legislation and, ultimately, to a state of siege in November 1902, when federal troops were called out to bolster the police presence. Curfew was established, and demonstrations were broken up. In October 1904 anarchists in Rosario, a growing industrial city northwest of Buenos Aires, called for a general strike that swept the nation, quickly paralyzed Buenos Aires, and left hundreds dead and wounded. Even the moderate middle class picked up arms in 1905, in an attempt by the fledgling Radical Party to unseat the ruling oligarchic government. Authorities imposed discipline in the streets, targeting anarchists—wildly popular for a time among workers—whom they correctly identified as the instigators of widespread revolt. Determined to maintain peace in the city's economic and political center, the federal government in this period turned regularly to repressive measures, calling out troops to maintain order and instituting states of siege, which imposed harsh, if temporary, limitations on the ability of city residents to move freely throughout the city, to demonstrate, and to publish subversive materials.

In one such event in the last months of 1907, the poor of the downtown area, who were fed up with the dramatic rise in housing costs, refused en masse to pay their rent. Women from the Plaza de Mayo, and eventually from as far north as Palermo, organized the strike, urging 100,000 of their neighbors to join the protest. Known as the Tenant Strike, it was the largest major collective action of the decade. Daily demonstrations of men, women, and children, the last carrying brooms "to sweep away the landlords," demanded their rents be reduced by a third. Landlords responded with mass evictions; local judges and district police stepped in to carry out the removals. In San Telmo, the historic downtown district, a teenager was killed and a number of other people were wounded, as the two sides faced off with pistols. A mass outcry at the boy's funeral prompted police to bring out larger numbers of enforcers: armed officers were ordered to guard judges and attend future evictions, standing in a line in the street with their long shotguns in hand. By early 1908, the tenants capitulated in the face of state violence and the potential for more. A few months later, city residents remarked that living conditions had, if anything, worsened in the aftermath of the strike.[36]

The 1907 Tenant Strike, May Day celebrations, and other labor demonstrations punctuated the city's routines on a monthly, sometimes weekly, basis. The daily papers reflected elite views of alarm about the rising social conflict and mass action. *La Nación*, the voice of the establishment, reported that toward evening one night in January 1912 a group of about 1,000 striking taxi drivers gathered in the Montes de Oca Street, in the immigrant neighborhood of La Boca. Several shots were fired into the air, with the goal, according to the paper, of "engag[ing] the police in order to excite the spirits of the workers and provoke a fight."[37] The weekly magazine *Caras y Caretas* (Faces and masks) took a more bemused approach to the strikes that were erupting on a nearly daily basis throughout the city. Reporting in 1904 on a tailors' strike, it included photos of tailors filling the streets around the Italian opera house, crowded in by horse-drawn carriages and watched over by police in helmets. The magazine's comments on the event made mild fun of Argentina's half-steps toward progress as being constantly mitigated by the tensions of the age: "Just as we were getting used to making ourselves fashionable, in gossamer little outfits with fanciful designs and completely new cuts, up went the flag of the tailors' strike."[38]

The capital police increasingly focused their attention not just on individuals but on groups of men in public. Observation of strikes, demonstrations, and other disturbances took on a scientific cast as the police documented and compiled statistics. A 1909 police report to the interior minister paid homage to the repressive 1902 Residence Law for reducing the need to expel "those persons whose background and behavior constitute a very grave danger to the public order." The police proposed expansion of this "law of public health" to control vagrants and other quasi-criminal social disturbances."[39]

A Science of Political Policing

While social pathologists formulated prescriptions, police bureaucrats sought to repress and control immigrant labor unrest and unruly crowds. Theoretical descriptions and abstract classifications were not enough. The mission of the police, according to Luis M. Doyenhard, chief of police in La Plata at the start of the twentieth century, was to respond to the dominant problems of the age: "social, political, and economic instability."[40] Antonio Ballvé, director of the Penitenciaría Nacional, former police functionary, and author of numerous studies of Argentine crime, hoped police methodology would keep his prison population down. He saw a new role for police in prevention and proposed a modernization and rationalization of police procedures. Ballvé wrote in his 1907 *Text on Police Instruction*: "The police agent should never forget that the most important of his missions" is fighting crime "through the most skillful, severe, and vigilant means."[41]

As the Argentine state attempted to professionalize their forces of social control, the Buenos Aires police grew in power. The year 1880 had marked not just the centralization of the nation but also the federalization of the Buenos Aires police. With its location in the capital, the Buenos Aires police gained strength and power from the national consolidation of the government. State leaders turned to the Buenos Aires police force to fill quasi-national policing functions, and the next twenty years saw the rapid professionalization and expansion of the police both in the capital and the Province of Buenos Aires, with the elimination of civilian posts and establishment of police schools. As the police amassed considerable institutional power, they drew closer to the military, which was itself undergoing a rapid expansion and consolidation of political power. In 1880 the capital police had an annual budget of 113,990 pesos; by 1902, the budget was over 5.5 million pesos.[42]

An official report suggested that a "School of Scientific Policing" should be set up to cope with the myriad and growing forms of crime. Such an institution was seen as "more urgent every day because of our demographic density and cosmopolitanism, the ample freedom to enter the country, the wealth of our environment, the enormous activity of industry and commerce and its respective parasitism." The school would train officers in the latest police methods, procedures, and technical skills, and students would learn about "studies of criminal anthropology, sociology, psychology, and legal medicine."[43] The *Archivos*, reviewing the book *Scientific Police* by the Italian criminologist Alfredo Niceforo in 1907, stated, "Among the new sciences, without a doubt, police studies are the most modern . . . of the current sciences."[44] The *Revista de Policía* was required reading for all police employees after its initiation in 1888.

La Prensa supported the plans to modernize and rationalize the police force, and it argued, given the rising crime rates, that the public, "with good reason," also demanded "good policing."[45] Its editors agreed with Buenos Aires police chief Luis Dellepiane that officers should have higher salaries, in order to "make available select personnel, educated [and] capable of preventing [rather than] fomenting street riots, so frequent and so upsetting to all."[46]

The left-wing and workers' press took an opposite stance, calling for reform of the police so that "its administrators understand the uselessness of the irritating and stupid conduct" of the police force. These reformers pointed to police corruption and abuses of power, and they labeled "police crimes—a public shame." The socialist paper *La Vanguardia* (The vanguard) recounted, with some satisfaction, that a "majority" of police officers (and even some judges) in the capital and interior cities were on trial for assault, murder, rape, and horse rustling. "Argentine Police—In Which Hands Is Social 'Order' Entrusted!" they exclaimed in 1909, and ended, with tongue in cheek, "Does this mean that all the police have been recruited to the jails"?[47] The feminist anarchist newspaper *La Voz de la Mujer* (The woman's voice), published intermittently in 1896 and 1897, had gone even further. In an article entitled "Pigs!" it responded to a report of a labor dispute in *La Vanguardia* (a "bourgeois socialist-pig newspaper") and concluded that the police were in the hands of shop owners who had ordered the arrest of a worker who had incited his companions to strike. "The crime was invented by the boss," *La Voz de la Mujer* protested.[48]

In "We Defend Ourselves," *La Vanguardia* referred to "daily combat" on the streets, with the police siding with the bourgeoisie against workers' groups. "The Police Is the Sole Promoter of Disorder," warned a headline of 1905. To *La Vanguardia*, "social order" meant free demonstrations of workers' rights. It published eyewitness accounts of beatings, unfair arrests, and maltreatment of those arrested and those awaiting trial. "Thieves are treated with more consideration by the police" because they were not perceived by the police as "dangerous enemies of social order like the striking workers." When a 1907 strike of dock workers was broken up by the police, the paper reported that workers were gathering daily in the Iris Theater, and spilling out on the patio, with an additional crowd waiting outside in the street to enter. With the two surrounding streets filled with "five or six thousand striking workers," an "enormous quantity," the police ordered their dispersal for lack of a permit to assemble. The police had "openly violated" the workers' rights to assembly, *La Vanguardia* wrote, and had "committed an intolerable abuse, and if tomorrow there was some extreme case, the police would be the only ones responsible, the only provocateurs."[49]

After 1891, police observation had taken on a "scientific" cast, with the tallying of strikes, demonstrations, and other disturbances. The capital police began to categorize its incident reports by national origin, counting arrestees from the largest immigrant groups, which reflected the then common ideas that linked Spaniards, Italians, and Jews with criminal tendencies.[50] Labor activism itself was classified as a criminal activity, and in the late 1890s anarchist violence and terrorism were inserted into a category of "crime against the social order." In 1895 police reportage on strikes and anarchist activity took up much of its annual report: "Where one could discover a true motive of alarm to the social order, it is in the anarchist theories, [which] have multiplied among many of the workers by means of active propaganda by European agents, whose appearance demonstrates clearly that this public danger threatens to grow."[51] In 1900 the capital police deplored the "notable increase in European anarchists who arrive on our shores, due to the absolute liberty that they enjoy here."[52]

Annual police reports to the interior minister also provided analysis of crowd behavior and working-class psychology. In 1904, a year of "numerous strikes," the Buenos Aires police department's annual report assured that "the repressive methods that we are obligated to adopt in certain circumstances have been limited to those strictly necessary to deter aggression and reestablish public order." The police saw themselves as mediators between employers and workers, as an interested party in the "social question" of the day. They suggested legislation to protect workers' conditions and help prevent the outbreaks of violent protest. Finally, they urged more studies of this "urgent problem," like those underway by hygienists and social scientists Juan Bialet Massé, Carlos Bunge, and José Ingenieros.[53] A few years later, in 1907, with accelerating concerns about anarchism, police chief Ramón Falcón wrote to Interior Minister Marco Avellaneda, "Sectarian propaganda and the deeds of the professional agitator that periodically disturb the public order affect in particular free labor and the regular functioning of the economic activity of the country." The next year Falcón advised Avellaneda, "Anarchist propaganda, with frankly ruinous purposes for the social body, is used practically on a daily basis by means of its publicity organs and various presses, with clear incitement to consummate all types of attacks, taking off from the sectarian aphorism 'property is theft,' advising plunder is all forms and . . . preaching crime, especially against the high mandates of the country."[54]

Passage of the Social Defense Law, an anti-immigrant measure, in 1910 provided for expanded police powers on the street and censorship of the left-wing press. Eloy Udabe, the chief of the Public Order Division, complained that with many anarchists in hiding, "the difficulties of observation and vigilance of elements that form the anarchist sect have grown considerably." Udabe proposed that the government create a special office to conduct background checks and identify criminals who were attempting to migrate to Argentina. Surveillance of suspect individuals increased significantly. Photographic records of criminals and suspects more than doubled between 1909 and 1910, from 22,146 in 1909 to 48,045 in 1910. Fingerprint records showed a similar rise.[55] By 1912, the annual police report noted, thirteen suspect immigrants had been arrested, eight of them deported. Still, despite the legal intervention backed by a stronger police force, there had been 102 strikes, "with a total loss of 3,523,116.30 pesos."[56]

"Our Police Have Obtained a Complete Success": Fingerprinting the Masses

Police hoped that new techniques of identification could be used in concert with the latest classifications of criminal types to register members of groups who were most likely to commit crimes and, if possible, to prevent their antisocial acts. Police triumphs in catching perpetrators, chief of police Alberto Capdevila wrote in 1887, were due to "advances in science [that] permit the notable improvement in the police services." Anthropometric identification, "like that in the Paris police," permitted the identification of criminals "through scientific procedures."[57] Fingerprinting was one of the most promising tools to control the movements and behavior of dangerous criminals on the street—prostitutes, repeat offenders, and anarchists. It spoke to the new needs of the non-intimate, nonlocal society that Argentina had become. Police had first used a latent fingerprint as forensic evidence in an 1892 double child murder case in a small town in the Province of Buenos Aires, but it was as a tool of identification for criminals and potential criminals that the technology took off.[58]

Both forensic and identification uses of fingerprinting were pioneered in the laboratory of the Buenos Aires provincial police, located in La Plata, where Juan Vucetich, a local police scientist, developed a classification and filing system for fingerprint records. Vucetich, who had recently emigrated from Croatia, focused on how to compare prints from a crime scene with ever-expanding individual identification records. His system was recognized by experts in Europe and the United States as a true innovation. The genius of dactyloscopy, as he called it, was not its accuracy (detailed measurements of finger marks were by then common), but the efficiency with which his classification system could be married to an emerging bureaucratic archive of individual prints. The ability to classify and retrieve large, if not unlimited, numbers of record cards had many potential applications beyond solving crime especially civil uses, like the maintenance of general population records, prostitutes' registers, and immigrant tracking systems. The provincial police began to fingerprint its own candidates, the first experiment in the civil use of fingerprints in Argentina. Of 373 candidates in the first year, eleven with previous police records were found.[59]

The publication in 1904 of Vucetich's *Dactiloscopía comparativa* (Comparative dactyloscopy) marked the perfection of his system and methodically laid out his science and his views toward applying it.[60] Since the key to his system was its ability to classify easily, the identity card created for each prisoner after the institution of fingerprinting in La Plata contained a series of descriptions based on anthropometry, such as size of head, size of foot, shape of nose, and color of skin, and imprints from all ten fingers. After fingerprinting, police officials would analyze and type each print, record the five-digit code on an identification card, which also held a photograph or other personal information, then file the cards by their print type in multidrawer cabinets, each drawer clearly labeled. Vucetich proclaimed that "today, all the well-organized police departments on earth use [fingerprints]. . . . The Province of Buenos Aires was the first of them. . . . Our police have obtained a complete success."[61] Vucetich stressed that his system was the most modern for "educated societies," because in a highly mobile and urban industrial world, repeat offenders harmed public security, making it "necessary to look for means of more exact and rational identification." He reported that his system had been instituted throughout the province, and that it "functions regularly, and secures the recognition of repeat offenders to justice, as it assures the elimination of individuals with bad pasts who want to join the police."[62]

Virtually all officials in the capital and province had embraced dactyloscopy by the mid-1890s; early skepticism within the ranks of the police had been swept away within a few years. The capital police's Office of Identification documented about 600 people in its first year, and by 1899, it had 11,322 identification cards in its cabinets.[63] In 1894 Bibiano S. Torres, in a pamphlet aimed at convincing police chiefs around the country to use the Vucetich system of identification, wrote of dactyloscopy that "if the police . . . has as its mission the maintenance of the public order, liberty, individual property and security . . . we have proposed to bring to the government's awareness an improvement that we consider a very necessary complement to police service." Scientific policing, Torres claimed, was an effective tool in the fight against the "association of thieves and assassins which, with refined cynicism, penetrates the most aristocratic centers of towns and cities."[64]

In 1901 Buenos Aires city police commissioner of investigation José Rossi attended a lecture by Vucetich on dactyloscopy and reported in detail what he had learned about the differences between Vucetich's and French rival Alphonse Bertillon's approaches to identification, as well as the theory of arches and loops behind Vucetich's system. Rossi's description indicated the relative simplicity and ease of the Argentine fingerprint system. He recommended its general adoption as "an acquisition of a new universal language, simple and efficient," and two years later, he introduced the Vucetich system to the Buenos Aires city police and mandated the adoption of dactyloscopy.[65] Police officials in the capital, in celebrated cases, exchanged prisoner fingerprint files by telegram to preempt jailbreaks.[66] In 1902 the provincial Supreme Court required judges in criminal proceedings to receive a report on each defendant from the Office of Identification, and it suggested that the legislature consider altering the penal code to specify harsher sentences for repeat offenders, many of whom would be identified as such with the use of fingerprint record cards.[67]

Antonio Ballvé wrote that "scientific observation has demonstrated that finger and palm impressions" have "an extraordinary importance for criminal investigations, if they are properly made use of," and José Rossi, who by 1907 had been promoted to city chief of police, called Vucetich's work "one of the greatest discoveries in applied policing in recent years."[68] Rossi's investigative division in the city police hoped to utilize dactyloscopy to carry out its mission to maintain public order, "by means of knowledge, preventative observation, and repression of socially disruptive elements"—in other words, to expand the identification system to the noncriminal population.[69] In 1916 alone, the capital city police took 141,691 fingerprints, bringing the total to 619,553 records (of a total city population of about 1.5 million in 1916).[70] Fingerprinting, the most rapidly applied criminal identification system and typology, provided a vital link between scientific generalizations of group behavior and the needs of the state to find and identify individual offenders and then to more easily identify suspect persons.

But the "complete success" claimed by police was belied by worsening violence and chaos in the public sphere. Efforts to bring the streets into order, seen as necessary to promote progress and industry for the nation, were hindered by the realities of a city bursting with new immigrants. There were too many people in Buenos Aires, with its crowded streets and tenements. The demonstrations and clashes of May Day 1909, and the warnings that the 1910 centennial celebration of Argentine independence would be disrupted, filled officials with near panic, which brought out an even greater willingness to use force. A growing right-wing vigilantism was also encouraged by police and by the state. In 1910 mobs of masked youths destroyed the offices of *La Vanguardia* and other left-wing presses and attacked people on the street. In the most violent confrontations these forces specifically targeted Jewish neighborhoods and businesses, actions that reflected an ideology that was bolstered by the published work of the most prominent biologists and criminologists of the day.

Police response to the acute social conflict of the second decade of the century was increasingly repressive. On May Day 1909, police chief Falcón sent out hundreds of cavalry officers to disrupt the demonstrations and ordered his men to engage workers in face-to-face fighting. The result was five dead and forty wounded. The chief ordered the arrest of scores of anarchists, Russian immigrants in particular, and in retaliation workers called a general strike. A week later, after numerous violent confrontations between demonstrators and police, the federal government summoned the army and the police to suppress the strike. Six months later, Simón Radowisky, a Russian immigrant anarchist, murdered Falcón and his secretary, Alberto Lartigau, on their way back from a colleague's funeral. With mounting tension in anticipation of the 1910 celebrations of Argentina's centennial of independence, the state of siege was renewed, and groups of nativist and reactionary youths, siding with police, organized to silence and repress left-wing newspapers, organizations, and individuals. Violent social confrontations, fostering a sense of generalized chaos, accelerated. The anarchist group "Light of the Soldier" ran through the streets breaking storefront and streetcar windows and beating town workers in their path; they were met by vigilantes, police, and military.[71]

Police official Luis Dellepiane responded in a benevolent and paternalistic tone to the social clash in his 1910 annual report in a section entitled "Politics and the Social Question." "This government," he wrote, "looks with the greatest sympathy on the manner in which the proletarian class, like all other classes, struggles for its material and moral interests, and since the government represents all the collective interests and not the hegemony of one dominant class, it is disposed to equally guard with the greatest respect the rights of the workers, and the rights, no less respectable, of capital."[72] But national and city leaders reverted to the swift use of force, censorship, siege, and threatened deportation as the most efficient tactics of maintaining social control, opting for age-old tactics of violence and authoritarianism.

SEVEN

Places of Regeneration

Prison and Asylum as "Medicine for the Soul"

> Punishment is medicine for the soul . . . a rationally necessary means to assist the will of the citizen.
>
> —EUSEBIO GÓMEZ

Argentine physicians, by the beginning of the twentieth century, had unprecedented access to the bodies and minds of those whom the state had put behind bars—criminals, the mentally ill, unruly women, and juveniles. In a typical case of such access, prisoner Vicente Diego, who was incarcerated at the National Penitentiary for three years for petty theft, endured "repeated examinations." Such examinations were not intended to cure the sick or bad, but to treat perceived degenerative and genetic flaws on the national body. The report on Diego offered evidence of "hereditary degeneration" and a history of his parents' mental illness and hysteria. It painstakingly recorded Diego's anthropometric measurements, including the size of his skull. Dr. Helvio Fernández, the psychiatrist who would succeed criminologist José Ingenieros in 1913 as editor of the country's top psychiatry journal, *Revista de Criminología, Psiquiatría, y Medicina Legal* (until then, the *Archivos de Psiquiatría, Criminología, y Ciencias Afines*), recorded "numerous physical stigma of a degenerative character," like facial asymmetry and malformed ears. Fernández assessed Diego's "unstable, nervous, and impulsive" behavior and even his character. "His *work habits* are very few, making laziness his most outstanding characteristic. . . . This lack of will is the offspring of his hereditary degeneration . . . . [It] has made him inept under normal circumstances in the struggle for existence and has led him to the easy commission of criminal acts." Fernández, logically, prescribed work therapy, "sustained and methodized" tasks in the prison workshops. Solitary confinement was forbidden on the grounds that isolation had already sustained or exacerbated Diego's condition, and it encouraged vices like masturbation that would lead to further degeneration.[1]

Since colonial times Argentine society had provided institutions to remove dangerous individuals from its streets. Prisons were central to the society. As in other Spanish American cities, every town in Argentina had its prison in the main square along with the church and *cabildo*, or meeting hall. Before 1877, prison confinement in Argentina had been minimal, haphazard, and miserable. The confined often languished in police cells without any hope of trial. Justice was popularly perceived as arbitrary. The mid-nineteenth-century dictatorial regime of authoritarian strongman Juan Manuel de Rosas, while perhaps less violent (or "barbaric") than late nineteenth-century liberal critics would have it, had employed archaic methods of punishment, including public floggings and executions, forced military service, fines, and imprisonment. The few prisons that existed were considered holding pens for prisoners awaiting trial or transfer, not sites of education or rehabilitation. The lack of prisons meant that a great number of male offenders served their sentences in the army. In both prisons and in the military, offenders were subject to torturous punishments, ranging from shackling and placement in stocks to whipping.[2]

In 1877 lawyer Fermín Alsina, later a member of the governing council of the state of Corrientes and one of the earliest voices for the enlightened rehabilitation of prisoners in Argentina, wrote his university dissertation on the topic of penitentiary systems and how they should be applied in Argentina. He argued against harsh penalties and proposed education and work therapy. "Penitentiaries should be practical schools where the inmate learns morality, and not dens of iniquity and degradation," he contended. His plan for a therapeutic prison regime reflected contemporary views of criminals as carriers of disease. Prisons needed to go beyond the "segregation of this sick member" from society to "achieve his reform" through specialized discipline. The tool of choice was work, "the great regenerator of society. . . . The man dignified by work will return to the breast of society to be a useful member of it." Reflecting the optimism expressed by many modernizers for prison reform, Alsina wrote that "an educated nation with good habits of order and labor makes severe punishment unnecessary."[3]

Prison reformers saw their work as part of a national project of manifest destiny, and many equated prison advances with national progress. Like their colleagues in criminological theory, prison scientists understood their efforts to be part of an international evolutionary process of ever greater improvement. When, in 1904, the *Archivos* published the speech of Joaquín V. González, justice minister, which had been delivered at a ceremony that placed Antonio Ballvé as director of the National Penitentiary, an editors' note praised the government for progressive change and for "this official word, inspired in ideas cemented in modern science." Such moves would help bring Argentina up to par with the "most civilized countries." Criminologist and prison official Eusebio Gómez, too, praised the modern changes as the salvation of Argentine civilization, essential to creating the spirit of citizenship: "the disharmony that is born of [an offender's] disorder disturbs the harmony of the entire social organism."[4] The new findings in penitentiary science must be "adapt[ed] . . . to the most urgent needs of our country."[5] Dr. Rodolfo Rivarola, an important legal scholar, pointed to the prisons as protection against "the immigrant wave": "Argentine legal experts, without trying to be sociologists, through their observation of certain criminal forms that appear in European society, and taking into account the immigrant wave and the increasingly European character of [our] society . . . can predict anarchist crime and the white slave trade as deeds meriting particular consideration." The prisons would be weather vanes of social disorder.[6]

By the late nineteenth century, in the name of modernization and Argentina's health and progress, science and the state sought to segregate not only violent persons and criminals but vagrants, orphans, the mentally ill, wayward wives, and prostitutes. Specialized institutions were built to deal with the range of "degenerates," born criminals, and anarchists; women and minors also had their designated places of confinement. Criminals, by grade, and the mentally ill were separated from each other to avoid cross-contamination. The cause of their afflictions was seen, paradoxically, both as inherited and as environmentally caused. Although confinement was intended to protect society from harm, there was more: the social pathologists saw places of confinement as sites of regeneration, renewal, and healing, where the confined could be treated as the sick. If crime were a disease of the social body, then prisons were social hospitals.[7] In the institutional sphere of the prison, inmates were a captive audience, people who had been compelled to give up membership in the outer world to become temporary or permanent wards of the state, living under constant surveillance and discipline. Prescriptions for treatment were far-reaching and invasive. Some of the broken could be made whole again, ready to contribute to society in their assigned roles. Far from the old-fashioned punitive model, the new approach was to be modern and humane. Punishment was not abolished, however. Medical, educational, and labor programs simply added layers of control.

"Moral Orthopedics": Specialized Institutions for the Mentally Ill, Women, and Juveniles

To address the nation's lack of efficient prisons, Eusebio Gómez, director of the National Penitentiary, recommended more specialized correction buildings and the classification and separation of offenders by type. For example, he proposed separating the accused from the convicted, inmates by types of crime, minors from adults, and men from women. Women, certainly, needed their own reformatories, as did youths. Repeat offenders should be separated from occasional offenders who could be rehabilitated. Noting a disturbingly high number of alcoholics among the adult prison population, Eleodoro R. Giménez, then director of the medical corps at the city jail, called for government support for "reformatories for alcoholics," citing the models of the United States and Europe, "where the governments, convinced of their effectiveness, have lent their decided support."[8] Lack of segregation and specialization was a principal problem at the National Penitentiary, completed in 1878; its population was "a heterogeneous amalgam of criminals of all types and ages," with youths intermingling dangerously with hardened criminals, and "occasional criminals" with "professional ones who shared their pernicious influences." In the words of prison official Rafael Súnico, "all moral and regenerative sense" was in danger of disappearing.[9]

Penologists and criminologists believed that regardless of degree of adaptability, those confined could be controlled and managed scientifically. Rodolfo Rivarola, a prominent legal scholar who represented the Argentine government at the 1914 International Penitentiary Congress, claimed, "The criminal . . . is in reality a passive subject of planned sentencing." Similarly, Eusebio Gómez concluded that "whatever its defects, a prison sentence is, and will continue to be, the most effective recourse in the secular struggle against crime. And since we cannot abolish it, we try to enhance its effectiveness. New ethical concepts [and] new forms of crime will modify the character of penal function, but it will always endure because crime is eternal."[10]

The critical concept was individuation of punishment and treatment, as Argentine criminologists like José Ingenieros, founding editor of the *Ar-*

chivos, had advised. For Gómez, with his hands-on prison experience, any attempt to rehabilitate was hopeless without individualized programs. "Each criminal requires a distinct jail," he exclaimed, and without such, "prevention would be impossible. How [can we] exercise the moral orthopedics that prison reform implies, employing identical methods for subjects whose deformities (some congenital and some acquired) stem from such different causes?"[11] The mentally ill, Rivarola told the International Penitentiary Congress, must be separated from criminals, women from men, and children from adults, to protect noncriminals or the relatively innocent from "hardened" criminals and "degenerates" in a place that was neither jail nor mental asylum. The disciplinary regimen should be tailored for the best outcomes for each group and each person. Criminologists recommended the "creation of appropriate establishments for their observation and treatment, then for the reclusion of these degenerates; jail and mental asylum are not adequate places." The psychologically unstable could be carefully distinguished from "born criminals" by their level of consciousness of their crimes and their degree of responsibility.[12]

Indeed, the mentally ill presented a unique set of challenges. Most visible were the disruptive mentally ill who had been removed from public view and sent either to a hospital, to a prison, or to police holding cells. In the mid-nineteenth century the semiprivate Beneficent Society was put in charge of hospitals and asylums, overseeing a growing number of small and mostly rundown institutions that typically housed between 50 and 150 patients. Most of the mentally ill languished there for years in substandard conditions. In the 1880s the state took over the administration of most asylums from the Beneficent Society. Some psychiatrists recommended removing the mentally ill from the family environment, both to protect the family and to make treatment possible. (Moreover, they feared that some forms of mental illness were contagious.) Patrolmen picked up deranged individuals from the street, or desperate family members dropped them off at hospitals and police stations. But with the lack of regulation in the hospitals, prisons, or asylums and with unclear legal requirements for commitment, psychiatrists gained a growing reputation for wielding too much authority over the mentally ill.[13] An observation room in the police department, set up in 1899 for the examination of mentally ill people, was directed by the psychiatrist and criminologist José Ingenieros in one of his first jobs after medical school. Subjects off the street were examined there, and detailed reports of their physiological and psychological status were provided to state officials. Within a few years, this power of examination had expanded to other institutions, first to the city jail, the Cárcel de Encausados, which held adults and minors who had not yet been tried or sentenced, and then later to the National Penitentiary.

Physicians and men in the new specialization of psychiatry, increasingly viewing the mentally ill as dangerous, proposed ways to confine them in the few existing (and inadequate) hospitals, most of which had been built at midcentury, such as the Mercedes Hospital (for men) and the Women's Mental Asylum. Psychiatrists and reformers recommended that the state build more asylums and modernize those already in place. As the century turned, they favored more repressive and controlling practices, like preventative institutionalization, and argued that such practices should be mandated by law. Some further proposed immigration control for any incoming individuals who exhibited symptoms. José T. Borda, future director of the Mercedes Hospital, warned that immigrants were pouring into the mental hospitals. Only strict measures imposed by the state, backed up with funding, would ensure that the "offensive" and at least potentially dangerous mentally ill could be controlled.[14]

The mentally ill were, for the first time, now linked directly to tendencies to crime and violence, with criminals and the mentally ill viewed as the same type of degenerate at simply different points on the continuum of decay. Alejandro Korn, who would later become director of the Buenos Aires provincial asylum, wrote in his 1883 thesis in medicine, "From a philosophical point of view, we can agree that there is but a difference in degree between crime and madness."[15] Similarly, in his book *Simulación de la locura* (Simulation of madness) Ingenieros wrote that criminality and madness were "branches on the common trunk of psycho-physical degeneration," of a "closely related parentage." Psychiatrists sometimes gave a mental patient no diagnosis more specific than "degenerate."[16] Social pathologists, comparing Argentina to nations that had achieved a degree of modernization, pointed out that it had a higher rate of mentally ill than nations like England but a lower rate of hospitalization; in 1919, according to Ingenieros, only 8,000 of Argentina's 15,000 "lunatics" were safely and appropriately institutionalized.[17] In 1906 the *Revista Penitenciaría* (Penitentiary review) had recommended a widespread study of mental illness in the prisons, considering it a humanitarian project, since the insane tended to have longer jail stays than others and could benefit from special treatment.[18]

Just as the vulnerable mentally ill needed to be housed apart from hardened criminals, so too confined women should be kept apart from men. Argentine colonial law had mandated that, to protect their "decency," women be incarcerated in separate prisons from men. Although a women's institution was built in the late seventeenth century, it was too small to hold more than a few prisoners. As a result, for most of the colonial era women were confined in a special section of the male prison. In 1877, when the new National Penitentiary was completed, men were sent there, and the older jail was converted to the female-only Women's Correctional House, overseen by the Catholic Order of the Good Shepherd. (Not until the 1970s did female corrections institutions come under state control, despite the secularization trend in the late 1880s.) About 100 women could be housed in the Women's Correctional House, but in any given year between 400 and 500 women and 1,000 and 2,000 girls passed through, sometimes confined for a few months, sometimes longer. Between 1890 and 1923, the institution received 18,629 prisoners and 34,623 abandoned minors convicted of no crime at all. (Abandoned infants were sent to the Casa de Expósitos, the city orphanage run by the Beneficent Society. Most girls over the age of six who had no parents—homeless, abandoned, or orphaned—were farmed out as domestic servants in private homes or sent to the women's prison.)

It was seen as an added advantage that, thanks to the nuns' assuming control for women and girls, the state would have minimal expenses, a fraction of the costs of running male prisons. Equally important, men were regarded as inappropriate authority figures for such women. Nuns were seen as ideal for this job since they were already accustomed to a cloistered life and spartan living conditions. The day for inmates in the Women's Correctional House began and ended with prayers, with other hours filled by cleaning and recreation and mass, confession, and catechism. Elementary education was provided for the poor and immigrant women. Women's chores were limited to sewing, embroidery, washing, ironing, and such, all designed to return the inmates to a more productive and orderly domestic state. The nuns tracked former inmates, measuring their success by the number who went on to create "Christian" families or, alternatively, to join the order itself.[19] Critics of the system, however, objected that the Women's Correctional House was old, decrepit, and crowded and that the nuns were incapable of and unwilling to apply "scientific" methods of rehabilitation. Abandoned minors were living with experienced female criminals, and critics considered the prison a "breeding place for prostitutes" and recidivists.[20] Even while the Women's Correctional House was under church control, however, state physicians did have limited access to the inmates.

The physicians and criminologists who examined women, much in the same way as they did men, prescribed different treatments, since women were considered less likely to commit crimes and, even when convicted, were less likely to be a social danger. According to the 1906 Buenos Aires prison census, only 3 percent of the city's inmates were women, and of those nearly half were serving sentences for property crimes; the most prevalent forms of assault or injury, committed by only 1 to 3 percent of the female prison population, were for "women's" crimes such as abortion.[21] Most women prisoners were released within a few months on the condition that they pursue "domestic therapy" of housework and child care. Religious, state, and scientific officials alike were convinced that women were best rehabilitated under the watchful gaze of family members, especially a husband, a father, or a brother. But many women were sent to houses of deposit (discussed in chapter 5), the punitive, segregated semi-jails for any woman who had challenged the male authority of her household, or who had been convicted of "forgetting her sacred duties as wife," for maternal "crimes," for "irascible behavior" like speaking insolently to her husband, or for actual or suspected adultery. "Depositing" women, purportedly to protect their honor and to keep them from the more dangerous and corrupting environment of the women's prison, was a practice designed to protect or avenge the honor of an offended husband. Houses of deposit were seen as a source of shame, and most women dreaded and resisted depositing at least as much as incarceration.[22]

Minors in jail attracted special attention from reformers who were concerned to protect and mold the nation's youth, especially its boys. In 1880 the justice minister's annual report included a plea from the Department of Juvenile Correction for separate facilities for wayward youths, "completely independent from adult inmates," with "penitentiary discipline, primary instruction, and arts and skills" to "prevent the frequent criminality in minor vagabonds and abandoned" youths. The report lamented, "These poor wretches . . . inspire pity today and terror tomorrow."[23] While girls were dispatched to church institutions, abandoned and criminal boys were sent to reformatories, little more than holding cells, with few amenities. With space at a premium, some boys were housed in adult prisons, where, according to the 1910 city census, they "receive a better treatment than in their usual life, and discipline which, if it does not reclaim them, at least often sets them on the path of social regeneration."[24] The reformatory at Marcos Paz also served as an orphanage for male youths who had had no encounters with the law. It was organized into "families," though there was some concern that older boys would corrupt the younger ones by teaching them low-life ways. Similarly, the mixing of boys with adult men could result in the boys' corruption and criminal education. As Gómez put it, prisons grew "a species of greenhouse poisonous plants." A man who spent more than a few months there became a "gangrenous member of society." The director of a prison must "apply knowledge of hygiene" to "impede contact between the professional wrongdoer and the newcomer and make impossible the propagation of the physiological and moral contagious germs that swarm in the atmosphere of this sinister city."[25]

Similarly vulnerable to corruption were inmates accused of crimes but not yet convicted, who were routinely grouped with "hardened criminals." In 1905 José Luis Duffy, head of the boys' reformatory, reported that, in response to long-standing "popular demands" from public officials and even from the president of the republic for a jail designated for persons accused and awaiting trial, the boys' asylum was to be converted into a jail, the Cárcel de Encausados, a change that would help bring the prison system in line with "scientific principles." Until that moment "the law only saw crime, applying abstract penalties, without occupying itself with the criminal."[26] In his 1906 survey of the nation's major prisons, Gómez noted that the Cárcel de Encausados, hastily established, had been created to segregate accused from hardened criminals. The new institution was "a true novelty, signifying at the same time an advance of inestimable value." It practiced advanced approaches, with medical exams, work therapy, and education for the adults and juveniles housed there. It was to replace confinement in the holding cells in police commissaries, in which, Gómez wrote, "one observes the harmful influence of the common prison. The convicts housed there, *a mountain of them*, in unhygienic *rooms*, pass the day meeting, conversing about their motives, praising or condemning the abilities of their defenders, and bitterly condemning the actions of the judges."[27]

Duffy described the "careful study" of adult and juvenile criminals by prison medical staff as the prison's principal means of "scientific organization." He cited one Dr. C. M. Hicken's reported comments that "the base of administration of justice is the exact and precise knowledge of human nature, and this can only be acquired with the detailed study of embryology, psychology, anthropology, phylogeny, cellular pathology, etc." Citing Italian criminologist Raffaele Garofalo, Duffy contended that "jurists do not know their enemy, who is the criminal. To know him, it is necessary to have observed him thoroughly in the prisons, in the penitentiaries, in the places of banishment." True applied reform, however, would be a challenge since "in our universities they only study abstract criminal law, overlooking the auxiliary sciences." He supported the notion, citing proposed legislation on the provincial and national levels, to require medicolegal offices to be linked to courts.[28]

Gómez, too, suggested that the reformatory would create a laboratory for pedagogy (to teach, as he called it, the "science of childhood") to "investigate all that can contribute to the most complete knowledge of the child, in all the aspects and manifestations . . . of the child, normal or abnormal." On a visit to the juvenile reformatory at Marcos Paz in 1906, Gómez praised its devotion to manual arts (primarily agriculture) and education. "To avoid the diffusion of crime, to diminish the use of repressive functions, it is necessary to dedicate constant attention to the moral education of boys [niños] . . . in order to counter the apparently fatal and inevitable influence of heredity." Only education could offset the effects of bad heredity or bad parenting, both of which were understood to be the main causes of juvenile crime. Children were easier to rehabilitate than adults, as children were less susceptible to "anthropological" factors and more responsive to education and therapy. Treatment could "destroy the infectious germs absorbed in the atmosphere of immoral homes, to substitute the lack of education or correct the defects of a bad education. . . . [T]his same education . . . can be effective in destroying or tempering the influence of ancestral defects."[29]

Under Duffy's administration, an expanded medical service assessed the physical health of the boys and men in the Cárcel de Encausados, identified and separated criminal types, and determined courses of treatment. It checked for "contagious, epidemic diseases, and those that could affect the morals or discipline of the establishment," and physicians vaccinated inmates on arrival (for which diseases, official publications did not specify). An even more in-depth examination, required on entrance and exit from the prison, recorded a prisoner's state of health and constitution, "aptitude for work," and some conclusions about the effect of his prison experience. Examiners also noted a prisoner's weight on entrance and on exit— according to their records in 1905, most inmates gained at least one kilo a year. As with prostitutes at the city's medical dispensaries, inmates found to be diseased were forbidden to leave until treatment had been completed.[30]

A government decree of 6 June 1905 expanded the medical service in the Cárcel de Encausados, creating the Office of Medicolegal Studies, whose central task was to conduct medical examinations of prisoners. Juan Pablo Raffo, a prison physician, explained to Duffy, in a letter, that a new system of notebooks or "Cuadernos medico-psicológicos" had been set up to file and organize the reports, each an eighteen-page booklet with space for detailed data that judges and attorneys could retrieve, including anthropometrics, psychological profile, and hereditary background. "Elemental order in the preparation of reports and the study of criminals" was essential and so was an organized place with data on prisoners' "social environment, hereditary factors, and psychological examination." Even bed-wetting and nightmares were to be recorded and kept on file, as were "level of education; scientific and religious knowledge; knowledge of manual arts and the laws and codes of the country; the changes in instinct, from fundamental and hereditary perversion to simple acquired disturbance; the knowledge of diseases and injuries from infancy; the evolution of childhood; the existence or not of syphilis, tuberculosis, and scrofula; involuntary urination, night terrors, [and] use of alcoholic beverages." Such information would "contribute to objectify the organic indolence of the subject, making his psyche trans-parent, so that the experts can form a more solid judgment, and result in a more exact appraisal of responsibility in the eyes of the law."[31]

A newly established journal, *Revista Penitenciaría*, would publish the medical reports "to shed some light on questions" of criminal law, applying "to practical cases the new scientific norms . . . [to see if] they obey the impulses of heredity, environment and education, factors that form [the convicted prisoner's] will and determine his decision." The government's official published record, the *Orden del Día*, included psychological exams and called for supplementary reports from the warden of the prison school and the chaplain, with each prisoner's social class, behavior, and morals to be recorded regularly. The Cárcel de Encausados staff aimed for nothing less than to "study the soul of our juvenile delinquents, and know their objec-tives, prejudices, and ideas." Prison physicians would thus conduct an in-depth investigation of each person. The practice of examining and report-ing would "prepare the ground for a more easy application of the coming reform of criminal law, especially if it sanctions probation." The reports were expected to assist in classifying prisoners and also to assist in demon-strating the necessity for specialized prisons.[32] Moreover, to inspire and teach the boys, "it is necessary to know well the character of the boys, and to . . . gain the confidence of each individual, stimulating his good inspira-tion, comforting him in his moments of weakness." Models of moral educa-tion were of "capital importance for the future of nations."[33]

At the reformatory, guidelines demanded that "all youths upon entering will be examined by the doctor, who will report respective to his physical aptitude for work."[34] Examinations of youth at the Cárcel de Encausados included interviews with them, their parents, and their teachers. The ge-netic and behavior "antecedents" of each prisoner were probed, to look for

clues to nervous diseases, venereal disease, and alcoholism and for information about parents' work habits and nationality. Physical exam reports
included detailed anthropometric data, like size and shape of ears, fingers, and facial and cranial measurements. Tests measured mental capacity,
memory, perception, and intelligence. Finally, the "moral sense" of each
prisoner was assessed as a guide to capacity for responsibility. In 1908
Eleodoro Giménez sent Duffy a statistical summary of 248 cases from the
prison medical office's first three years of activity. He wrote that the reports
written at the request of the court "demonstrate all the transcendental
importance that such studies have as an unshakeable base of scientific truth,
on which not only the imposition of penalty on the criminal should be
supported, but also the investigation of his origins." In a number of cases,
judicial error—especially on responsibility for a crime—had been prevented
by the timely introduction of "scientific" reports on defendants stricken
with illness, alcoholism, or other afflictions that diminished their responsibility. "It is not enough," Giménez wrote, "that there exists in the conscious of all the profound conviction that no defendant should be condemned without a previous medicolegal report. It is necessary to elevate it
to the terrain of practice, that it be incorporated into our criminal legislation, that it comes to life in our judicial procedures as a means of humanity
and progress." The medicolegal reports were a critical step to "procure the
means of our social perfection."[35]

"A System of Rational Separation": The National Penitentiary

Back in 1869, the provincial government of Buenos Aires had mandated the
construction of a new penitentiary, intended as the first of its kind in Latin
America.[36] The competition for architectural plans required proposals that
segregated residents by sex and that separated convicted prisoners from
those awaiting trials and adults from minors. At its maximum, the penitentiary was to hold 600 men and 100 women. The government specified the
height of a wall to separate the prison from the surrounding streets and
warned the engineers to "abstain from proposing anything to do with pure
architectural adornment, considering that they should not try to construct a
building of luxury"; moreover, planners were to "propose the project in the
most complete manner and in accordance with the advances of science and
experience that have been proven in other lands."[37] With the construction
completed in 1877 and the prison under national control in 1880, Buenos
Aires criminals, as well as some from provinces with no "secure prisons,"

were housed there. The justice minister, who oversaw the prison system, had taken as his model foreign states, especially Britain and the United States, that had succeeded in adequately funding and building modern penitentiaries. In the 1880 annual report of the prison, Edmundo O'Gorman, the administrator in charge, described "the double point of view of the moral improvement of the criminal as well as the guarantee of order and security to which the Community aspires and deserves."[38]

According to the director of the National Penitentiary, Antonio Ballvé, the institution was founded under enlightened principles: the government mandate had reflected the "influence of the new ideas about punishment that would clear paths, dispelling classical prejudices and conquering the resistance of custom." In drafting the institution's layout and regimen, officials had been inspired by the penitentiaries of Philadelphia and Auburn, New York. The National Penitentiary's floor plan followed that of the famous London Pentonville prison, and the disciplinary regime that of the penitentiary in Auburn, New York—work and education by day, silence in solitary confinement by night. Though intended for convicted criminals only, because of lack of facilities in the city it actually served as a holding cell for hundreds of detainees awaiting trail. It had five pavilions, each with one washroom, one lavatory, and 120 cells, which were 4 meters long, $2\frac{1}{3}$ meters wide, and $3\frac{1}{2}$ meters high ("sufficient capacity for each individual"), and two pavilions for women with 104 cells. There were multiple workshops, gardens, patios, and an infirmary, all with "perfect" ventilation and modern running water and gas lighting.[39]

If the penitentiary were used solely for prisoners convicted of "serious crimes," Ballvé reported, the number of cells would be sufficient, but it was not—with too few appropriate general jails, the penitentiary was crowded with between 900 and 1,000 inhabitants, not 700, a significant number of whom were petty criminals and the accused awaiting trial. It was a shame that Buenos Aires had inadequate prisons for a "city of its importance." However, "in practice," Ballvé said, "the prisoners are put into a system of rational separation that counteracts the indescribable dangerousness of prison life. The background of each prisoner, the class of crime that he is accused of, his age, education and social position, are all factors that determine the division into groups, and that in particular secure the separation of repeat offenders or professional criminals from the occasional or first-time offenders."[40]

The middle-class weekly magazine *Caras y Caretas* (Faces and masks), in a 1907 article entitled "Modern Prisons," praised the comforts of model

penitentiaries in Argentina and abroad. It pointed to a new general infirmary, then being built, with "all the conditions of a true hospital, with
six halls, minimum capacity of 60 beds and maximum of 84, and all the
necessary conveniences to join in it all the sanitary services of the establishment: consulting rooms, operating theaters, pharmacy, chemical and
bacteriological laboratory, odontological cabinet, electrotherapy, [and] ra-
diography." According to *Caras y Caretas*, comfortable facilities led to the
"surprising [and] complete transformation of the characters of certain types
of criminals in whom change seemed impossible." Indeed, at the National
Penitentiary "the convicts' luck to pay for their crimes in prison increases
daily. Banished from civilization are the torturous procedures that in ancient times were applied to criminals. They are gradually acquiring a state of
well-being."[41]

By contrast, most Argentine provincial prisons had been held over from
colonial times with no renovations. The justice minister reported in 1881
that "with exception of the National Penitentiary, the prisons in our country, in contradiction to what the Constitution mandates, are not healthy, nor
clean, but rather sites of infection, dirt, crowding, [and] inhabitability, that
produce precisely that which our founding document wished to avoid: the
torture of our prisoners."[42] When Gómez visited the nation's prisons in
1906, including the jail in Sierra Chica, the Agricultural Reformatory in
Marcos Paz, and the National Penitentiary and the Cárcel de Encausados in
Buenos Aires, to assess how well they followed the tenets of criminal anthropology and new penological approaches, he offered conclusions that
were largely dismal, though the penitentiary itself was an exemplar of "scientific criminology" in action. The Sierra Chica jail, southwest of the capital
city, was "a true sewer that spills out into society a current, a permanent
source of infection and contagious germs, physiological and moral, that
poison, stupefy, depress and corrupt," and moreover, there were blatant
signs of rampant homosexuality, or "acquired pederasty." There was no
prison education or viable work program to speak of. Gómez even deplored
the color of the uniforms inmates wore, a bright red. Perhaps the worst
aspect, however, was the "lack of *treatment* that develops in [the prisoners]
the sentiment of their individual value, a sentiment that gives force to the
struggle and that determines in the damned the agreeable illusion of ex-
tinguishing his sentence to be able to enter in the social union [*consorcio*],
contributing his healthful efforts for his own benefit and for the general
good."[43] Gómez recommended a takeover by federal authorities, a nationalization of the prisons. He did not trust the provinces, with their backward,

undeveloped ways, to create the proper structures and apply the new approaches to prison discipline. Although the National Penitentiary in Buenos Aires was supposed to be a model for the nation, by 1900 there had been little, if any improvement in jails in the interior. A 1905 editorial in the inaugural issue *Revista Penitenciaría* noted that the science of prison organization in Argentina "finds itself in embryo."[44] An article from the socialist paper *La Vanguardia*, citing the experience of an "immense quantity of prisoners who exist in various jails" in La Plata, reported that prison experiences led to "repugnant promiscuity," rape and sodomy, and the exposure of minors and untried accused to "hardened" criminals. The paper put the blame squarely on the government.[45]

At the 1914 Argentine Penitentiary Congress in Buenos Aires, Eusebio Gómez warned that "an enormous prison population, lacking the most basic resources, conspires against the security of all. Disorder assumes such proportions that it causes national shame."[46] "The current state of the prisons of the province [of Buenos Aires]," said Gómez, "do not follow ideas of a scientific order, nor of culture, nor of humanity. The inmates live the life of animals, and chronic idleness characterizes their situation."[47] Francisco de Veyga argued that for habitual criminals the only "effective and legal" remedy was "definitive sequestration . . . the privation of civil rights, that is to say, a legal declaration of incapability." The state should assume "the guardianship that accompanies the suppression of capacity . . . as the only possible solution for the grave problem of habitual crime."[48]

Even the highly praised National Penitentiary, being overcrowded and out of date, could not control Argentina's burgeoning criminal population. In 1889, 663 prisoners were housed in 600 cells in the five men's pavilions. By 1897 there were 5,153 individuals in the same space. The workers' press, too, was skeptical about the "modernity" of the National Penitentiary and the treatment of the convict class. For example, *La Vanguardia* ran regular coverage of the arrests and incarceration of workers who were rounded up by police from the streets, and in an 1905 editorial, "Our Prisoners," the editors protested the sudden increase in "persecution of our *compañeros* [comrades]."[49] Commenting on appalling medical conditions for prisoners, the paper promised exposés of "how the poor prisoners, who have the misfortune to fall sick, are treated." Medical officers in the prison, the paper argued, cared less about the health of the convicts than about determining whether they were faking illness. The prison physician himself hardly ever showed up, evincing gross "administrative negligence." Seemingly, the attitude was, "If prisoners die, who cares!"[50]

"True Innovation in the Study of the Criminal":
The Criminology Institute

Joaquín González, a former minister of justice in charge of the Cárcel de Encausados, hoped to establish medicolegal examinations of the psychological, medical, and biological traits of persons confined as a key to prison reform and specialization. Such prison clinics would "bring about a true innovation in the study of the criminal."[51] The subsequent founding, in 1907, of the Criminology Institute at the National Penitentiary was a significant victory for Argentine criminologists, which signaled their influence and growing legitimacy and provided a central location for scientific research and publishing. The rising careers of professional criminologists, their involvement in government affairs, and their international reputations exemplified the status of the field itself. During the first two decades of the twentieth century, these social scientists lobbied hard for concessions like mandated medical and psychological exams of arrested persons, the creation of special institutions of confinement for particular groups, and the expansion of the penitentiary system into the nation's interior. Founding director José Ingenieros, introducing the institute in the *Archivos*, wrote that its goals included to prove "the characteristics of Argentine criminality, and to gather more effectively to solve our own problems of prevention and repression."[52]

Referred to within the penitentiary as the Psychology and Anthropology Office, the Criminology Institute was trumpeted by Argentines as the first site of clinical criminology applied to penitentiary science in Latin America, if not in the world. It was also viewed as a laboratory for testing and advancing European criminological theory and as a potential producer of scientific knowledge for the international community. Ingenieros described the tasks of the institute as "criminal etiology, clinical criminology, and criminal therapeutics," which reflected his new classificatory system of "psychopathology." Under "criminal etiology," the causes of crime, Ingenieros listed physiological and psychological factors (the study of criminal anthropology) and also sociological and environmental factors. He described "clinical criminology" as the scientific effort to "establish the degree of social inadaptability or individual dangerousness." The institute's "therapeutic" section would include "preventative institutions, legal applications, and penitentiary systems," and the data gathered from etiological studies would inform preventative legislation, just as the data from clinical criminology would guide the organization of penitentiary systems.[53]

The inauguration of the institute had been met with international fanfare. The Argentine government invited foreign scientists and journalists to attend its opening, and in 1907 and 1908 experts from France, Spain, and Italy visited the National Penitentiary and published their impressions in a range of European and Argentine journals. To the Argentines' great pleasure, the Italian experts who came included world-renowned figures such as Enrico Ferri, Guglielmo Ferrero, and Gina Lombroso, daughter of the famous Italian criminologist Cesare Lombroso. Gina Lombroso pointed out in the Rome journal *L'Avanti* that the Buenos Aires National Penitentiary was "established according to the recent doctrines of the Italian anthropological school," and she referred to Antonio Ballvé, then director of the National Penitentiary, as a "fervent disciple of Lombroso and Ferri"; Ballvé had "with great care observed all the precepts dictated by my father." New prisoners were studied "physically and morally, investigating their hereditary antecedents." Ferri praised the modernization of the "model penitentiary," noting its "considerably radical reforms," among them the publication of the *Archivos*, the reform of criminal laws, and the adherence to the Italian school of criminal anthropology.[54]

The Argentine prison innovators slavishly followed European models of criminology. As José Ingenieros wrote upon the inauguration of the Criminology Institute in 1907, "The modern criminological doctrines, which for a quarter of a century have been exposed, discussed, and corrected in the scientific world, have from today an ample sphere of study and application in the Argentine Republic. . . . It is a pleasure to note the significance of this new organism to the triumph of scientific tendencies which have updated classical penal law. . . . The work begun by illustrious criminologists and alienists . . . now begins to bear fruit."[55] Criminology Institute member Horacio Areco asserted, "I make a vote for the prosperous and fertile continuation of the work of Enrico Ferri, whose studies have had such influence on our scientific culture"—and here he listed sixteen Argentine specialists, including José María Ramos Mejía, Francisco de Veyga, Eusebio Gómez, Antonio Ballvé, and Juan Vucetich, as "the founders of these investigations in our country, preparing the data and criteria which will serve in the future to constitute Argentine criminological science."[56] Ballvé promoted Ingenieros as well suited to lead the institute because his international scientific reputation guaranteed that foreign collaborators would maintain interest in the Argentine prison system. Through the *Archivos*, "the results of [the institute's] work, such as all the documents, statistics, studies, etc. . . . will be published and made known in scientific centers and among people who are interested in this class of questions, both in and outside of the country."[57]

The Criminology Institute, like its predecessor, the laboratory at the Cárcel de Encausados, examined inmates and compiled dossiers, which it called *boletínes médico psicológico* (medico-psychological reports). Eusebio Gómez described the institute's methodology as an examination of "the predisposing and determining causes of the crime, the means of committing it, and the concurrent circumstances; the conduct of the subject during the process, his judgment of the crime and his sentence, his adaptation to prison life and the probability of reform." The Criminology Institute, Gómez said, classified each prisoner "according to his psychopathology . . . by the general classification of delinquents devised by Ingenieros." Prison officials assessed the offender's conduct three times a year in a report attached to the offender's medico-psychological report.[58]

At the center of the effort to modernize and rationalize prison treatment scientifically was the biopsychological medical examination, whose results were used to specify individualized "treatment" for each prisoner's regeneration and also to maintain internal prison discipline. This examination was designed to help modern judges, prosecutors, and defenders make wise decisions about treatment to protect society and avoid recidivism. In the words of Gómez, in a 1914 review of the inner workings of the penitentiary, "The first preoccupation, then, of the authorities at the [National] Penitentiary, is to know, as thoroughly as possible, the subject who will be subjected to treatment." The Criminology Institute was charged with tracking especially closely the disruptive elements within the prison population, "all the inmates who present signs of mental illness and all believed to be epileptic, alcoholic, or victims of any physio-psychological disturbance." Institute staff were to "intervene in all cases of suicide and deviant acts" and "advise the director of the prison concerning moral and intellectual education and labor of the inmates." They even kept an archive of inmate tattoos.[59]

The decision to locate the Criminology Institute at the National Penitentiary, with its diverse population of criminals, reflected a central goal of scientific penal reform: individualization of criminals and specialization of their treatment. These reports were considered official court documents, for use as evidence in trials.[60] Gómez explained that experts were required to examine each inmate as he entered the prison, in order "to know his background, [and] methodically apply the precepts of correctional pedagogy." They set up individual dossiers [*expedients*]; at the time of admission to the prison, police submitted folders [*prontuarios*] with data on the crime and personal data on the prisoner, such as his civil status and level of education. Next, "immediately after, comes the physical information about the criminal; his height, weight, skin color, hair color, beard, shape of the forehead, relative sign of the eyebrows and eyelids, color of the eyes, form of the nose, mouth, lips and ears, his register of individual fingerprints, a list in extensive detail of his distinguishing marks, all this accompanied by four photographs taken at distinct times." Knowing these traits, along with age and level of education, would help officials identify the proper level of treatment—which form of labor would be best, which emphasis in the program of moral education, and whether the prisoner should be put in isolation. Officials were to observe criminals and make elaborate "diagnoses" of their level of degeneration and their prognosis for rehabilitation. A panel of prison officials, drawn from the prison school and wardens, "will be charged with collecting data relative to each prisoner, to establish a classification of his conduct, taking as a basis that which [is observed] in the pavilion, in the workshop, and the school . . . and finally, the manifestations of his character, his tendencies, education, morality, and other particular circumstances that can serve to judge him." The dossiers would help determine treatment or punishment and, further, assist in scientific studies of criminality and its causes. The individual criminal body thus became the subject of science, and taken en masse, these bodies would shed light on "the study of . . . the most interesting problems that criminological science presents us."[61]

In implementing this process, Argentine criminologists and penologists believed they had advanced beyond their Italian colleagues, at least in some cases, such as that of the murderer Alejandro Puglia (see chapter 4). Originally misdiagnosed in his home country of Italy, Puglia was subjected to a more through examination once he was in the hands of the Argentine criminal psychiatrists. There they analyzed his past behavior, past evaluations, his present condition, and proposed remedies in order to establish whether Puglia was a "lunatic" or a malingerer, whether he was responsible for his crime or not, and what his degree of "dangerousness" was.[62] They concluded that he was criminally irresponsible and must be segregated not only from society at large but from the prison population. Though Puglia was faking illness, he was mentally deranged and should be committed permanently to a mental hospital, Ingenieros said: "Given the advanced and liberal criteria of our prison system, the presence of Puglia in the Penitentiary is poisonous, as it is to society. The prison work regimen—which is the basis

of all criminal discipline and reform, when that is possible—requires shar-
ing and the use of instruments, which in some circumstances could be used
as weapons by those who use them."[63]

Regeneration through the "Love of Work" and Civic Morals

With the anthropometric and psychological examinations, Argentine prison
officials could shift from overseeing barbaric and pointless punishments to
carrying out carefully calibrated individual treatments. National Penitentiary director Eusebio Gómez cited a "profound revolution" in modern
science's study of crime as a "biological and social phenomenon, and punishment as a function of the social organism." Punishment was now "inspired by other sentiments" than revenge. The "desire for vengeance had
been replaced by conservation; fears substituted by the desire for the good
of the species; the hate with which the criminal was treated in all time (to
the refinement of wickedness) came to be replaced by 'philanthropic treatment that recognizes at its base the functional and organic unity, and a
harmonious synthesis of all constitutive parts of the social nucleus.' The
prisons should not serve exclusively as a refuge for those who have broken
the law." Rather, they were a place of regeneration. "It is the duty of high
humanitarianism," Gómez contended, "to dedicate intelligent attention to
the transformation of the criminal consciousness, with an ample knowledge
of the psychophysical coefficients of the criminal." Though modern society's "multiple vices corrode the social organism . . . the truth is that a
penitentiary system founded on scientific and humanitarian bases will have
and has had healthy results."[64] A 1913 prison reform proposal similarly
reported, "It is not enough to simply not denigrate the prisoner. . . . [T]he
scientific requirements of our time demand something else. To defend itself
today from criminal acts, society detains its perpetrators, tries to reform
them through work and education [and to] restore them to a free life, which
is a source of order."[65]

A humanitarian prison science, such as at modern penitentiaries like U.S.
prisons Elmira and Sing Sing, models of "*yanqui* spirit," would help usher in
civilization to Argentina. Ballvé wrote in 1907: "Penal systems in civilized
nations evolve rapidly, animated by the spirit of our times. . . . Modern
prisons are losing the sinister character that used to distinguish them. . . .
Punishment has lost in both legislation and custom the ancient concept of
vengeance and today is considered only an instrument of indispensable
defense for social coexistence."[66] Gómez himself mused, "If punishment is

medicine for the soul," then it is a "rationally necessary means to assist the will of the citizen."[67] A later assessment in the 1914 *Revista de Criminología, Psiquiatría, y Medicina Legal* (Review of criminology, psychiatry, and legal medicine) stated, "We know that our prisons, which used to be instruments of physical and mental torture, are now healthy and clean, as demands the most advanced principles of modern penal science."[68] Indeed, the prison regime was so "soft and gentle," Ballvé complained, that "certain subjects with abnormal spirits, hardened professional criminals, congenital rebels, [and] degenerates . . . do not produce, perhaps, all the desired effects. Therefore this system needs, by way of a complement, either a rigorous prison in a remote spot, like that of Ushuaia in Tierra del Fuego . . . or a section of our own jail, constructed especially for absolute lockdown [*regimen cellular absoluta*], in which they can be maintained in a rigorous solitary confinement and under a special regimen."[69]

Prison remained a place for punishment, said Gómez, but in the context and service of rational and scientifically calibrated discipline: "Punishment would not satisfy the intended goals if it cannot impose true suffering on those whom it is intended for. In our century of anesthesia, it is said, moral cures cannot work without pain."[70] In fact, punitive discipline was often justified to protect the greater good: "Social defense calls for, undoubtedly, the use of coercive means."[71] Although coercion in the name of defending rational, modern state interests was an acceptable form of discipline, the key concept in Gómez's analysis was prevention. He wrote, "Punishment should be purely preventive . . . its regimen tending, whenever possible, to readapt the criminal to the conditions of [his] social environment."[72]

Gómez identified three major innovations in the treatment and rehabilitation of prisoners: highly organized and controlled discipline, education, and regimented work. To Gómez penitentiary action was "educative by definition," since it supposedly forced "respect for order and the constitutive authorities, and obedience to one's rulers"—all requirements of normal social life.[73] He described comprehensive control, in theory, over nearly every action of the prisoners as the core of the new regime. In a modern penitentiary, "the whole of its experiences and occupations . . . come together to guarantee the security of the prisoners, not through the thickness of the bars, nor through the severity and rigor of the guards, but rather through the appeal and authority of the interior regimen exercised on them." This power would come, naturally, as inmates internalized the discipline, developing "their mental and affective faculties, because they awaken or perfect their manual and artistic attitudes, discovering in this manner the first roads to rehabilitation."[74]

Cleanliness, order, and silence were key. So was strict hygiene, as Ballvé put it, "not only for reasons of health, but also for educational motives, because in many cases, to get a new prisoner to learn to be a clean man means to win the first battle against bad habits." Regular baths, modern sewage, and clean water were required in prisons. In his 1914 survey of the penitentiary, Gómez described prison hygiene as a "preoccupation," "an educative task, destined to benefit them [prisoners] morally and materially." On their arrival, all prisoners were subjected to five steps: a hygienic bath "at the temperature preferred by the inmate"; a medical exam; immunization; haircut and shave; and disinfection of clothing.[75] Thereafter, inmates were required to wash up twice a day at the sound of a bell and to take a daily bath. They were not allowed to wear a mustache or beard, and their hair had to be kept short. In describing the arrival of the inmates to the National Penitentiary, the *Statistical Yearbook of Buenos Aires* noted that in addition to a bath, medical examination, and issuing of a prison suit, there was an obligatory orientation to "disciplinary regulations," in which the inmate "is reminded that obedience, work, and silence are his main obligations."[76] Prisoners were allotted four hours for hygiene, meals, and brief rests; nine hours for work; two for education; one hour for homework in the cells; and eight hours for sleep. An ample diet, with meat and bread baked in the prison, was to be provided, and the health of the prisoners was "to be carefully watched."[77] (In a survey of the prison diet, Gómez stated that the "character can be strictly linked to a change in tissue," which in turn is affected by diet. He listed the daily ration of food as more than 4,000 calories per prisoner per day.)[78]

Similarly, in the Cárcel de Encausados, officials implemented a strict daily regime of limited visiting hours and minimal free time. Prisoners were allowed a pack of cigarettes a day, use of the library, three outside visitors per month, and two opportunities for weekly "free correspondence." Discipline included "three principal elements" for "moral regeneration": a disciplinary regimen, education, and work. Such a disciplinary system, Ballvé pointed out, was "severe, as corresponds to the nature of the institution, but at the same time it is humane and based in the most strict justice." Individual treatment "is the ideal." Full compliance, if not passivity, was demanded: "Absolute submission to the rules, imposed with the most rigor from the first moment on all prisoners, will assure the instilling in them of the conviction that while enduring his punishment, he should consider his will annulled." A board of prison administrators was to observe each prisoner and classify his behavior on a six-point scale, ranging from "dreadful [*pésima*]" at the bottom to "exemplary [*ejemplar*]" at the top. Each point corresponded to a series of imposed hardships or rewards. The balance of punishment and reward "stimulates the development of the prisoner's own moral forces, applied to the greater good [*al bien*]."[79]

Work was a key component of rehabilitation and regeneration—in Gómez's words, the "principal attribute of discipline." It was instrumental in establishing order in the prisons, and it reflected the "laws" of larger society, "the best guarantee of public order and general peace."[80] Gómez proclaimed, "Our penitentiary regimen looks also in its own time at the regeneration of the criminal, placing him in conditions in which he could be reincorporated into the society from which he was temporarily eliminated, [and] exorcize his helplessness, by dedicating himself to the art or occupation that he was taught during his time of confinement." As a result, Gómez wrote, "work in the prisons does not inspire in him the desire to aggravate the law, but rather, it is used as an instrument of regeneration, [and] at the same time as one of the most efficient resources for the maintenance of discipline." Work "should be an inexorable law in the prisons." To inculcate a work ethic the scientists relied on their interpretation of the social theory of Herbert Spencer, that "the process that brings the prisoner to acquire work habits, is a process corroborated by his own power, this being exactly what is required to arrive at being a good citizen."[81]

Not just any work would do. Work should be "useful and moralizing" but also remunerative, providing concrete rewards and a professional persona to prepare the prisoner for the "struggle for existence" outside prison walls and eventually to make him a productive member of society in "a new life of peaceful and honorable labor."[82] In one of the penitentiary's twenty-one workshops, a print shop that printed and bound criminological studies, inmate workers in 1913 earned a total of 1,422,261.89 pesos from sales of these books, of which 605,262.86 pesos went to the state as profit.[83] (Ballvé, in another context, pointed out that the production costs of the *Archivos* were minimal because it was produced by prison labor.)[84] Prisoners were paid for such work, but their pay was garnisheed for their upkeep, and if they had dependents, a portion was sent to their families.

Criminologist Cornelio Moyano Gacitúa wrote that "modern criminal science has demonstrated the injustice connected to prison sentences without work. . . . So that punishment is not a reward and that the prisoner does not improve his conditions in jail, he should also be subjected to the law of work."[85] Labor as therapy, especially for men, was the principal means to regeneration. Gómez declared, "Laziness is the mother of all vices." He offered as its antidote the "regulation of the psychic and organic life of the prisoner," where work was "an agent of moral therapy" and "imposed in a rational manner." Officials in the new prison system hoped to "stimulate . . . the love of work" in minors and adults alike, while combating the spread of dangerous social theories and the breeding of vice, laziness, and rebellion.[86] Through labor they hoped to prevent bad behavior, even bad thoughts, from taking root and growing among the prison population.

Gómez also recommended structured, focused education in the prisons to develop "sentiments of individual and social duty" and "expand intellectual abilities." Education could "reprogram" prisoners' base instincts. Citing French psychologist Gustave Le Bon, Gómez argued that proper education would strengthen "the spirit of observation, of decision, precision, and cooperative discipline."[87] Such enlightened forms of criminal rehabilitation would defend society: "The Constitution does not consider the prisoner as reprehensible, but as a wretched soul, vulnerable to reform and moral sanitation. The concept of punishment is based on the principle of social protection that suppresses the crime on grounds of security and not with the intention of harmful punishment."[88]

The educational program at the National Penitentiary was obligatory and consisted of "reading and writing, Spanish [*idioma nacional*], morals, and history," math, science, and manual arts (penmanship, drawing, horticulture, and typing). The prison school, founded in 1886, employed a director and fifteen, "all normally trained" teachers; it had a library of 1,600 volumes, which was heavily used by the inmates. (By 1914, the number of volumes had doubled, and inmates were reported to have read 15,079 books the previous year.) To counter "internal barbarism" the new prison administrators offered an explicit code of morality.[89] Such a moral education did include religion, since "religious sentiment is undoubtedly one of the chords that vibrates intensely and endures in the criminal soul; it makes sense, then, to utilize this factor as a regenerative element, without prejudice or dogmatism, which are anachronisms in our time."[90] Nevertheless, religion, for men at least, was suspect. It could lead men to irrational behavior, even madness, or indicate a tendency for old-fashioned values. And in the prisons, it was "dangerous and extremely censurable." Gómez made a clear distinction between religion and morals: "They have nothing in common," he wrote. Even though they incorporated religion into their rehabilitation program, the prisons' pedagogical approach had to be scientific. "Respect for order and established authority, obedience to leaders are social necessities independent of all religion."[91]

A 1913 editorial in the *Revista de Criminología, Psiquiatría, y Medicina Legal* similarly underscored the prisons' emphasis on morality: "Moral responsibility and moral education of the will are the two principal points of the system." Indoctrination of youths in religion could "avoid extremist socialist theories and anarchist ideas that penetrate the soul of the prisoner."[92] Gómez suggested that the ideal education inculcated patriotism, the highest order of spiritual sentiment, perhaps even replacing worship of God. As director of the penitentiary, Gómez set patriotism as his goal in dealing with a multinational, heterogeneous prison population in need of unity and higher common goals.

PART IV *Hygiene*

EIGHT

Public Hygiene against Foreign Contagion
and "Sanitary Anarchy"

The nation's health is the supreme law of the State.
—NATIONAL DEPARTMENT OF HYGIENE MOTTO

A visitor to Buenos Aires in 1880 would have arrived in a city deep in the throes of modernization and construction, fueled by booming trade and population growth. The laying of pavement, sewer pipes, and streetcar lines, the erection of new buildings and a new port—all of these were going up seemingly overnight. The city virtually buzzed with activity. Within a few decades, Buenos Aires had a completely different atmosphere. Unadorned buildings from the Spanish colonial period were outnumbered by buildings in modern French and Italian styles. European-style sidewalk cafés lined the streets. Avenues modeled after French boulevards were wide and shaded by trees. A bustling luxury shopping street, the *calle Florida*, was surrounded by the elegant downtown homes of many of the nation's elite. New theaters and an opera offered some of the world's finest performances. The city had transformed itself from a "large village" to the "Paris of South America."[1]

Buenos Aires's remarkable transformation was made possible through dramatic investment in sanitation. As late as the 1880s, after heavy rains the sewer system would back up, leaving streets and even homes flooded with human waste. The city's inhabitants dumped waste and household trash in the river, which regularly flooded neighborhoods near the port. Unevenly paved streets and sidewalks created mud banks after rain. Adding to the general contamination were slaughterhouses and cemeteries, whose decaying and infected matter spread throughout the city. Hygienists, declaring the nation's health "the supreme law of the State," now spurred the government to invest in sewers, running water, garbage removal, and vaccination programs, virtually all in the capital city, leaving the rest of the country in a state of primitive sanitation.[2]

Like the real Paris, Buenos Aires, too, had an underbelly of poverty and misery, which elites sought to minimize, contain, or sanitize. New public health institutions went hand in hand, at least in theory, with expanded powers of surveillance and inspection. In 1893 Argentina's National Department of Hygiene issued regulations charging sanitary inspectors with checking buildings, ports, and ships in the capital and, eventually throughout the country. Sanitary police, trained inspectors who were to work closely with the city's police, were hired.[3] With the newfound capital from exports and additional funds provided by foreign (mostly British) investors with a stake in Argentine development, officials in the newly federalized city made major public works a priority. These reformers, led by Torcuato de Alvear, who served as the head of the city's municipal council throughout most of the 1880s, hoped modernization and beautification would attract investment, increase trade, and boost the national wealth. They focused on the aesthetics of the city at the same time as they built sewers and hospitals and paved streets.

Sewers and drainage were central to the nation's modernization goals. State officials measured the nation's health by rates of disease, ideas, and morals, and by control of dangerous germs, dangerous behavior, and dangerous ideas. They invested in disease control and bacteriology and used a bureaucracy to discipline foreigners and the poor through sanitary inspections, medical exams, and fingerprinting. The government marshaled hygiene as a civilizing and modernizing force, a way to engineer social change. In its view, degeneration in its many forms—diseases of the nation's "germ plasm," mental disorders, subversive ideas—were contagious. While reformers' efforts were viewed as part of a benevolent and liberal agenda, the rigorous intrusions in people's lives and ordering of behavior led to an obsessive control, especially of immigrants' bodies, characterized by humiliating removals from apartments in "unhealthy" neighborhoods, invasive physical exams, and the selective use of fingerprinting. The fixation on cleanliness would be used to justify a nascent bureaucratic authoritarianism. Moreover, the influx of immigrants gave state policy a distinct xenophobia, which equated immigrants with disease and dirt.

The proposals for reform, and the accompanying legislative and bureaucratic practices, were rampant with pseudoscientific ideas. Central to the debate and the governmental programs were concepts of contagion linked to racial inferiority. Only the right type of European immigrant would build

Argentina into a civilized nation. Attempts to rid Argentina of "barbarism" were, essentially, attempts at racial engineering, "a merciless war against the degeneration that threatens our future humanity," in the words of Argentine eugenicist José Angulo.[4] The program had three goals: cleansing, selecting, and purging. It called for policing the nation's borders over which disease, both metaphorical and real, might pass, and for curing the pathologies of the existing population. A broad set of long-term policies and agendas accompanied prescriptions for short-term "cures" in private homes, on the streets, and in institutions introduced by the social pathologists. Those who did not or could not conform faced possible violent removal from Argentine society or forcible containment within it. To carry out these policies, and to provide justifications for them, state officials turned to the social pathologists and to the most authoritative language of the day: science.

Public Hygiene as a "Material Religion"

Throughout the north Atlantic world, not Argentina alone, states that were armed with the emerging facts of hygiene and bacteriology sought to institute cleanliness. No realm was spared: home, school, and workplace. Books on personal grooming, clean food, hand washing, and the importance of neat and pressed clothes appeared. Legislation regulated the inspection of foods and factories. "Cleanliness is next to godliness" was the motto of the era.[5] In the 1870s a new awareness of germ theory and a belief that disease could be prevented with the proper hygienic measures spread among the nation's leadership in Buenos Aires. Bacteriology promised, so hygienist Luis Agote said, to "take a big step toward the solution of one of the greatest problems in humanity's interest: the defense against exotic migrating diseases." Agote, a physician and bureaucrat who would later find world fame for a blood transfusion technique, urged "civilized nations" to "introduce in their sanitary laws modifications. . . . The application of scientific truth, daughter of observation and study, constitutes the greatest advance in a country's laws."[6]

. This crusading approach was a response to the high toll of unsanitary conditions in an accelerating urban, industrialized society. Between 1872 and 1906, epidemic diseases were responsible for 46 percent of all deaths in Buenos Aires.[7] The city was, hygienist José Penna said, "in a perpetual epidemic state."[8] Yellow fever, which broke out during the stifling summer months, could kill at its peak hundreds of people in a day; in the particularly bad epidemic of 1871, it took 10 percent of the city's population. Cholera waves broke out every decade after 1850, with devastating scourges in 1886–87 and 1894–95.[9]

Hygienists also responded to the prevalence of tuberculosis, a virulent cause of death. In 1899 Samuel Gache, a prominent physician and social commentator, urged his peers at the Círculo Médico Argentino to mount a more concerted campaign against the disease. In 1901 Gache, Emilio Coni, and a number of other prominent hygienists organized the Argentine Anti-Tuberculosis League, which was funded by the interior minister and private donors, and modeled after similar institutions in Europe and the United States. In its first four years, the league opened two 500-bed hospitals, two 300-bed sanitariums, and twelve dispensaries. Its members also pushed for quarantine and disinfection of tuberculosis hospitals, and they proposed laws to require sanitation in schools and the pasteurization of milk. A 1902 municipal ordinance for the "General Prophylaxis of Tuberculosis" required infected persons to report within twenty-four hours of diagnosis. Physicians, landlords, and occupants of a patients' house or building were obligated to report new cases. After "denunciation" a house could be disinfected by employees of the Sanitary Administration; this procedure had been on the books since the 1880s for other contagious diseases but was unevenly enforced. Residents of poor neighborhoods, the exclusive target of these laws, were most often visited. In addition, the ordinance could call for forced hospitalizations and quarantine. Gabriela Laperrière de Coni, wife of the prominent hygienist Emilio Coni and an influential reformer, cited six main factors as predisposing workers, especially women (considered the weaker sex), to tuberculosis: alcoholism, unhealthy living conditions, poor nutrition, overwork, a lack of hygiene, and other "excesses." She and other reformers proposed as remedies labor reform and social reforms in housing and nutrition.[10]

Control of venereal disease and prostitution was also a top public health concern. Legislation regulating the space and free movement of prostitutes gave physicians broad authority over slum dwellers and other "quasi-criminal" social groups.[11] In 1907 Emilio Coni formed the Argentine Society for Moral and Sanitary Prophylaxis, the first concerted effort to combat both tuberculosis and venereal disease. Dispensaries were set up, linked to antituberculosis clinics. Coni's proposal for sex education, however, encountered fierce church opposition. While his plan did not get support from the state, it served as a model for the essentially similar Argentine Social Prophylaxis League, founded in 1920.

Hygienists were concerned, too, with the sorry state of Buenos Aires hospitals. Still, hygienists emphasized the importance of keeping ill and mentally ill people off the streets. In the view of Guillermo Rawson, societies had to choose between two approaches, either "the poor receiv[ing] in their homes the assistance of science" or the creation of special institutions—hospitals—to provide treatment; of the two, the latter was the more efficient means to achieve the goals of public health. But the hospitals of Buenos Aires were a horror. Rawson noted the great amount of waste and human fluids that fell to the floor in the Men's General Hospital, spreading odor and germs and sometimes catching fire. Floors, walls, bedding, and the layout of latrines—everything in the hospitals—must be precisely ordered, easy to clean and sanitize.[12]

The church opposed most of the hygienists' projects, understanding that such plans were part of a general secularization effort. To reduce the death rate from contagious diseases, state reformers sought to take over church-run institutions, such as cemeteries and marriage, and to impose regulations intended to prevent venereal disease. Eduardo Wilde, a physician who served as president of Argentina in the 1870s, proposed to raise the minimum marriage age from twelve to fifteen years for girls and from fourteen to eighteen for boys. Hygienists also sought an increased role for the state in coping with poverty and disease, instead of leaving such problems to charities to resolve. In doing so selectively, however, they abandoned most women's and children's needs to the church (see chapter 5).

Argentina had set up a national "medical corps" back in 1852, bringing together the capital's small Escuela de Medicina (medical school), which was by 1880 a full medical university at the University of Buenos Aires, the Hygiene Council, and the Academy of Medicine to promote sanitary controls and the regulation of pharmaceuticals. After 1867, the corps gained the power to impose fines and penalties for infractions.[13] But the city of Buenos Aires itself did not establish its own public hygiene council until 1871, which was converted into the National Department of Hygiene after federalization of the capital in 1880. The council was led by a series of distinguished physicians, among them Luis Agote, Emilio Coni, Eduardo Wilde, and José Ramos Mejía, all pioneering hygienists who were concerned with living conditions and the environment but also with behavior and mental states as causes of disease. By 1874, the medical school was incorporated into the University of Buenos Aires, and after 1885 the groundbreaking Institute of Legal Medicine was established.[14]

Two powerful liberal politicians who had been trained in medicine, Guillermo Rawson and Eduardo Wilde, embodied the close alliance of medical and state interests. Rawson's 1876 book, *Conferencias sobre higiene pública* (Lectures on public hygiene), stressed the importance of the environment—climate, air quality, water quality, and electricity—and provided an ideal layout of streets in urban centers as "the lungs with which a city breathes."[15] "In a narrow street, the air does not easily circulate and the light and health penetrate only with difficulty," he wrote. Streets should be paved to cut down on dirt and stones and stagnant water. There must be regular garbage removal. In 1880 his survey of findings on slum conditions was published as *Report on Tenements*.[16]

Links between the university and the public health apparatus were close. In 1875 Wilde, then a member of the faculty of the Department of Legal Medicine, joined the two-year-old Department of Hygiene at the university and, with Rawson, taught the first class in public sanitation. Wilde served as interior minister under President Bartolomé Mitre, and later as president of the nation. In 1880 he was a member of the city's Commissions on Running Water and Works of Sanitation.[17] He also helped implement laws to regulate and promote public hygiene. Rawson, one of the earliest pioneers in the city's sanitation efforts, was the son of an elite family from San Juan province; his father was a North American physician. Rawson served as interior minister and in the Senate in the 1850s and 1860s, and in 1873 he was appointed the first professor in the University of Buenos Aires's new Department of Public Hygiene, which he had helped to establish. "To whom," he asked, "is the study of Public Hygiene relevant?" His answer: "To everyone; young and old; women and men; lawyers and physicians; the worker and the man of letters. . . . Hygiene should be the material religion in all lands." Rawson drew attention to the narrow and crowded streets, the few safe public spaces, pileups of garbage, insufficient potable water, and ancient latrines. The prescription for Argentina's public health included university training in hygiene for medical professionals and sanitary inspectors, building and improving hospitals, and enforcement of modern scientific knowledge about, for instance, street sanitation and proper burial of the dead.[18]

Most immigrants moved out to poor yet tidy suburbs, but the most visible working-class housing units in the city were crowded one-story tenements called *conventillos*. The city's middle and upper classes thought these decaying homes, which the well-to-do had abandoned (but still owned), were an eyesore, as well as sources of contagion, of pestilence, and of immorality.[19] Landlords yielded handsome rental profits on these properties, with as many as 350 individuals living in a space previously occupied by a family of 25, including servants. The *conventillos* jammed families in single rooms, and some families even took in boarders of their own. Groups of six or more men might share space. The residents of the inner-city slums were about 50 percent of the city's populace in this period. Census takers estimated that between 66 and 72 percent of them were foreign-born. Municipal and federal authorities targeted *conventillos* as the source of epidemic contagion during the outbreaks of yellow fever. In 1875, in response, the city government required the inspection of dwellings with more than four rooms.[20] Inspectors were authorized to evict tenants from rooms of pestilence, depositing their occupants on the street.

By 1892, when the government had reorganized its archaic system of municipal inspections under the city's Sanitary Administration, more regular, more thorough, and more efficient official visits to poor housing were instituted, with "hygienic inspections" of homes every two months. The inspectors were intended to "assure the immediate compliance of all the prevailing ordinances about hygiene, morality, and security" and to provide written documentation of inspection results.[21] The city sought to sanitize homes of native-born as well as immigrant workers. The Public Assistance Office tried to intern people with cholera, but their families resisted, believing hospitals to be dangerous. When poor living conditions and rapidly rising rents prompted the slum tenants to rise up en masse, most dramatically during the 1907 Tenant Strike, violent shows of force from the police and fire departments backed up the landlords.

Zealous public health administrators intended to use force to carry out the measures they had succeeded in passing. In 1887 top hygienist Emilio Coni lauded the Department of Hygiene as being critical to the "considerable progress" of sanitary organization. The department regulated medical and pharmaceutical practices, performed sanitary inspections, and advised local governments of "lacunae in public hygiene" and national authorities "in cases of consultation and as court experts in legal medical affairs." Coni urged all provincial and municipal agencies to follow the Department of Hygiene's example in compilation of statistics and creation of a medical police corps to oversee inspections and vaccinations. Coni was especially impressed with the new "special health corps" that was now policing Argentina's ports and inspecting entering foreign vessels. All members of this special corps had to hold Argentine citizenship; all were required to be medical doctors with specialized training in "medical geography," "exotic contagious diseases," and "naval hygiene."[22]

Coni's 1891 reference text for physicians and lawyers, *Código de higiene y medicina legal de la República Argentina para uso de médicos, abogados, farmacéuticos, etc.* (Code of hygiene and legal medicine in Argentina for the use of physicians, lawyers, pharmacists, etc.) was nearly 700 pages long and included four major sections: the code of hygiene, a list of laws on public health; the legal-medical code, a summary of foreign and Argentine legislation and of procedures in medical jurisprudence; a code of medical ethics (based on the American Medical Association model); and a medical guide, with practical information for physicians. There were also lengthy discussions of criminal responsibility for alcoholism, insanity, abortion and infanticide, and suicide.[23]

The 1894 "Law of Prophylaxis against Infectious and Contagious Diseases" required that "the police and other authorities will lend to the authorities charged with carrying out this Law, assistance through public force as it is required." A few years later the municipal government went even further, creating specialized police task forces within the Department of Hygiene to deal with specific sanitation problems. Mandated by a series of laws passed in 1897, a corps was set up to police the ports; another for factory inspections and "industrial policing"; another to serve as a "mortuary police," to enforce proper disposal of cadavers; and a corps to deal with animals.[24]

José María Ramos Mejía, who headed the Department of Hygiene from 1893 to 1899 and became widely recognized for his contributions to the sanitary infrastructure of the nation, was praised in the professional newspaper *Semana Médica* for his progressive and nationalistic program, enacted to "control the border, but also improve the sanitary status of the country, organize the distinct and corresponding services, providing them with sanitary elements, train ad hoc personnel, put sanitary works into place in the provincial capitals . . . study the medical geography of the nation, investigate scientific needs to know the prevailing diseases among livestock, [and] establish sanitary codes, the national code." The newspaper commended Ramos Mejía's creation of the Sanitary Administration and the National Hygiene Institute and his consolidation of the sanitary legislation "necessary to end the ruling anarchy in this area, and to forever destroy quackery [*curandrismo*] with its disastrous effects on public health."[25]

The National Department of Hygiene attempted to deal systematically with "hygienic" problems such as living and working conditions among Buenos Aires's poor. The findings of its study were published in a 1910 volume optimistically entitled *Las conquistas de la higiene social* (The conquests of social hygiene), which described the advances achieved in European counties and what Argentina had to do to catch up. Augusto Bunge, author of the report, a department physician and member of a prominent Buenos Aires family, recommended legislation to regulate work hours, sanitary and safety inspections, and laws to protect women and children.[26]

In 1913 José Penna, in a letter to the interior minister, reminded him that "the country needs to maintain its reputation for health" and supported efforts of public health reformers to convince the federal government to allocate 3 million pesos to programs to prevent infectious disease. Law number 7444 had already mandated establishing sanitary stations in the nation's ports and major cities, with twenty-six stations devoted either to "maritime sanitation" or "internal prophylaxis." But by 1916 Penna lamented that this project was still not realized, admonishing the government, in his *Atlas sanitario argentino* (Argentine sanitary atlas), that "hygiene has to be the science and practice of strong, great, and progressive nations."[27]

The government had been lulled, by the early 1900s, by a prevalent sense that the worst of the epidemics were behind them. Officials had achieved basic sanitation in the capital and reasonable success with vaccination. A 1904 municipal census had reported that the epidemics were "a menace that will not return." (In reality, large, underserviced pockets of the city, especially in the houses that had sprung up along the unpaved streets of the suburbs, were suffering reprisals of epidemic disease.)[28] With sanitation more or less under control, hygienists turned their forces to "social plagues" like venereal disease, alcoholism, and tuberculosis, and to the working conditions and health of workers, women, and their children. Internal sources of contagion—hazards understood as both biological and moral—threatened to destabilize the public order and derail the nation's progress. Ideological and behavioral pathogens would require different kinds of sanitation from the control of contagious germs to the control of contagious ideas. Argentina planned to deal with this full range with the collaboration of its liberal modernizing scientists. Officials at the Department of Hygiene, understanding antisocial behavior as a public health problem, thus published reports and analyses, not only of communicable disease, but of crime and, even, of anarchism, in *Anales del Departamento Nacional de Higiene* (The annals of the national Department of Hygiene). Hygienists, financed by an activist state, demanded legislative measures and creation of welfare agencies to address such social plagues. A broad-scale solution would be required.[29]

"Selective Immigration with Scientific Criteria": A Solution to "Deplorable Ethnic Conditions"

The Madero port, carved out of the mud banks of the Rio de la Plata in Buenos Aires, had opened in 1889 to great fanfare. To President Carlos Pellegrini, the opening of the port was both a real step in building Argentina's economic power and a symbol of Buenos Aires's transformation from sleepy village to vital metropolitan center. The passage of people and goods, each with separate arrival areas, would, President Pellegrini announced, "offer the greatest symbol of human fraternity, linking the labor of all nations, to realize the common good."[30]

With immigration breaking the 100,000 mark for the first time in 1884, and by 1889 up to 289,014, the port's completion was essential to facilitate the processing of these masses of people.[31] But the influx presented new dangers as well. In 1898 a renowned physician, Luis Agote, then secretary of the national Department of Hygiene, warned that "the splendid ships, pride of modern naval science, charged with transporting goods from all over the world, are, by disgraceful contrast, the most frequent vehicle of said germs."[32] Teams of physicians were dispatched to board arriving ships to check for signs of disease, dirty clothes, germs, and filth and to carry out medical examinations of immigrants in the reception area and in a newly constructed Immigrants' Hotel (Hotel de Inmigrantes), just north of the piers.

Increasing the population had been a founding principle of the nation since the 1850s, and until the 1880s immigration seemed uncomplicated and was uncontroversial: Argentina's mandate was inclusive, with very few limits on immigration imposed. Though on a different scale, Argentina shared an inclusive ethos with other immigrant countries like the United States, with which Argentine state officials often compared themselves.[33] The General Department of Immigration was formed in 1856 by business interests, with funds and a building from the federal government; its goal was to protect and foster immigration. In 1869 President Domingo Faustino Sarmiento created by decree a Central Commission of Immigration, charged with providing a unified immigration service. With immigrants considered "indispensable" for building the nation's infrastructure and for participating in the national project, "to share with us the dangers of the struggle for liberty and institutions," even a small decline in the number of immigrants was seen as alarming. In the 1870s the commission recommended policy reforms and sent laws to Congress to set minimum health and berth standards for immigrant ships, among other measures. It also claimed the power to inspect ships and requested a new immigrant asylum to replace the dilapidated existing one (built in 1857). The government, the commission said, should draw up a legal definition of the immigrant that would exclude "undesirables" like the old, the sick, and the criminal.[34]

In 1876 the commissioner general of immigration, discussing the competition among receiving countries for the "upstanding" immigrant, noted that Argentina seemed to have a reputation abroad for failing to protect "security of life and property."[35] As an example, the commission's report cited a British newspaper article that claimed that Argentina had a high murder rate and little punishment for offenders. Argentine officials feared that publication of such negative information or rumors would steer desirable immigrants toward other countries instead, such as the United States and Australia, and steer infected or defective immigrants toward their own. Miguel Cané, an influential legislator, argued in 1899, and again in 1902, for more restrictive laws against immigrants "thrown from Europe who expulses them like a poison threatening its organism."[36]

As immigration accelerated through the end of the century and the city grew more crowded—and foreign—epidemic diseases seemed ever more exotic. Immigrants became identified as carriers also of moral contagion— prostitution and pimping, crime, destabilizing political ideas. The problems of the immigrant influx had been experienced by other immigrant countries but seemed particularly acute for Argentina. Interior minister Manuel Gálvez pointed out in 1910 that because Argentina was a large country with many open entry points, medical inspection and border control were more difficult.[37] Yet Argentina's economic growth depended on large numbers of newcomers. The state and its scientific allies strove simultaneously to encourage waves of immigrants and to select from among them. Argentine elites, perceiving the nation to be engulfed by a tidal wave of southern and eastern Europeans, tried to build dikes to prevent the nation from becoming a victim of its own success. The government hoped to manage the large number entering the country by disciplining newcomers, by racially engineering the nation to minimize the degenerative effects of immigration, and by literally transforming the immigrant population in terms of scientific criteria.

Argentina, like many other liberal democracies that inspected and detained foreigners and imposed medical inspections, followed a model of medical health as a forerunner to social and political health. State officials at various levels of government called on physicians and public health innovators to help select and assimilate immigrants. In this program of control of an influx of contagious germs, behavior, and ideas, "hygiene" encompassed more than medicine. Its bureaucratic champions and practitioners at the legislative level relied more on eugenical ideas, after 1900 just starting to take hold in Latin America. The efforts to select healthy citizens upon their entering through the port of Buenos Aires reflected the first moves to incorporate such ideas and practices into state structures.

Federal authorities, terrified that Argentina's open gates would admit new "toxins," dangerous to the national body, hired public health physicians, psychiatrists, and criminologists as gatekeepers to identify desirable and undesirable traits of the potential citizenry and to filter out symptoms of "barbarism." Santiago Vaca-Guzmán, a Bolivian diplomat who lived in Argentina, wrote that "the foreign element, as an active power, will purify and invigorate the nation's political forces." But he also divided immigrants into two groups: the first, the hardworking peasants who intended to work the land, and the second, those who would "swarm" in the cities, "lacking industry [and] moral energy, degraded by poverty and disposed to move to whichever part of the world [they] believe to find easy means of living without having to work." Such undesirables harm "new nations who have taken them to their breast because they consume without producing; they do not know how to earn their bread." The problem, said Vaca-Guzmán, was that Argentine law did not distinguish between the two types.[38] The weekly magazine *Caras y Caretas* (Faces and masks) reflected a positive view of urban immigration, displaying in its New Year's issue of 1914 a line drawing of a strapping immigrant, with chiseled features and muscular forearms, standing proud in a small sailboat, a hammer thrown over his shoulder, with the industrial skyline in the background. The caption read, "Immigration as a means of progress and culture."[39]

If the orderly flow of people was the fuel of the modernization process, the new generation of state leaders would make a concerted effort to achieve it. Political elites worked hand in hand with the emerging medical community, seeing science and public health as the most promising, and most acceptable, approach to the immigrant question. The program to select for the proper attributes of the new population expanded rapidly within a few years, mobilizing public health bureaucrats, police, municipal officials, and national legislators. Modernizers strove to merge the state with public health, hoping its practitioners would carry out key tasks of social policing. While immigration officials were secular and scientific, ultimately their methods served the moral ideology of the state. The medical inspections embedded political and economic agendas as well. The real goal of the sanitary inspections was to communicate ideals not only of health, cleanliness, and industriousness but also of racial and spiritual unity.

As the most important gatekeepers, the sanitary inspectors who boarded the ships had to decide "whether the character [of an observed illness] is lasting or dangerous." An 1875 report stated that "one sees the danger and the grave responsibility of the doctor who must be vigilant and arrange the measures taken to avoid the propagation of the case."[40] Inspectors had to make quick decisions whether to send an individual to the hospital or to order quarantine or to exclude him or her from entry. Officials recognized the need for new sanitation laws and practices, and a new awareness of germ theory and prevention of disease through proper hygienic measures spread through Buenos Aires. Agote called for the end of "sanitary anarchy" by regulating and modernizing the dangers of immigration in a progressively more mobile world.[41] Ideas about selection were codified as early as the 1876 Immigration Law, whose third article provided that the state "protect immigration that is honorable and industrious, and devise means to contain the current that is defective or unfit [viciosa ó inútil]." Article 32 elaborated what sort of individuals would be excluded: "those with contagious diseases or organic defect that makes them unable to work; the demented; and beggars, convicts or criminals."[42]

Physicians were thus called upon to define the desirable traits on which selection of immigrants should be based and also to inspect the new arrivals. Juan Alsina, an immigration service physician who contributed to discussions of immigration in the nation's census reports, said in 1899 that the law had, as a first priority, regulated the passage of immigrants since 1876. Ship captains were required to report any persons with "diseases of epidemic or contagious character" (yellow fever and cholera being the greatest fears) and to prevent any such persons from disembarking. "Immediately after its arrival in the Republic, [a ship] carrying immigrants will be visited by a group composed of the sanitary doctor and an employee of the Immigration Office with the objective of investigating the sanitary status of the ship," Alsina advised.[43] "The constant arrival of ships from all nations and points of origin," he continued, "brings to our port a great quantity of individuals who can carry with them or in their baggage or merchandise the germs of epidemic ills." He recommended that immigration officials, as in the United States, introduce a system of medical certification, to be validated by a port physician before disembarkation: "In this way, one makes sure that those individuals who enter the country are healthy and one avoids the danger of diseases and hospital costs. In the Immigration Hospitals [in the United States] they only assist immigrants affected by transitory diseases, those with chronic diseases are thrown out of the country, without exception." The same system should be introduced in Argentina and the medical corps expanded to manage it: "The State expense will be less than the copious sums spent on the maintenance of Hospitals, in which a large number of those treated are individuals with a small amount of time resident in the country and who brought their disease from another country."[44]

Luis Agote, too, lamented that the horrible conditions on passenger ships meant public health problems for Argentina. He described "the poor hygienic conditions in which those unhappy ones travel to come to our country in search of fortune, elements who are exhausted by poverty and work." Those conditions "result in *a travesty that lasts more than three weeks, in tiny infectious places, without air, without light*." He pointed out that one ship, the *La Plata*, which had originated in Italy, arrived in port with forty-five cases of diphtheria. The city had no means of quarantining such large numbers of ill arrivals. In addition, it had to deal with the more commonplace "totally debilitated immigrants, without strength and with repugnant parasites on their bodies."[45]

In the 1890s the national treasury designated new funds to construct an Immigrants' Hotel and for new commissions; the total budget for immigration nearly doubled from 536,000 pesos in 1880 to over 1 million pesos in 1899.[46] By 1903, all passengers in second- and third-class passage were by law "visited and meticulously inspected by a group composed of the Immigration Inspector, Sanitary Doctor, and Official of the Maritime Prefecture, who verify the hygienic and health conditions of the boat. . . . All this in defense of the immigrants."[47] Funds were also provided to tabulate vital statistics on the newcomers. Immigrants had been counted since 1857, but now they would be described meticulously by nationality, level of education, physical health, and visible stigma. A monthly chart for 1889 shows a breakdown of ill immigrants by nationality, the majority of whom were Spaniards, followed by French nationals, Belgians, British, Italians, and Russians. The ill were tabulated also by age and sex and origin of illness. An 1890 publication from the Immigration Department reported that "the final tally [for 1890] reveals one ill person out of 28, of the 135,670 immigrants; one death for every 324 of these immigrants; and one death for every 11 ill people. These numbers are truly extraordinary. Never has the number of immigrants entering the country been elevated to this degree, and this excess is the consequence of the overcrowding of people on the ships, that with the food and treatment, leaves much to be desired in these cases and explains the large number of sick people." Moreover, widespread illness was not only attributed to overcrowded ships, but to "low-quality immigration, above all among certain groups of Dutch, Belgian, and Spanish immigrants, who have been the most numerous this year."[48]

As the threshold of the new port was trampled by the 6 million people who had filed through by 1914, it was increasingly clear that the state could not inspect them all. The number of available sanitary inspectors never exceeded a few dozen. Issues of funding, organization, and personnel prevented a truly efficient and universal screening process. There was subterfuge, too, no doubt, on the part of some immigrants and patients themselves, who may have sought to avoid denial of access by trickery and deception. The medical inspections were cosmetic at best. They did not address all potential sources of contagion; for example, officials focused exclusively on the port and ignored immigrants arriving over land, who might also have brought pathogens with them.[49] Argentina's medical inspection of immigrants was a largely symbolic gesture, a harnessing by the state of the authoritative and legitimizing language and techniques of modern medicine, to transmit the image of a civilized nation to the multitudes crossing its threshold.

Comparing their policies with recent legislation in United States, Canada, and Australia, Argentina's legislators in 1909 demanded:

> The liberalism of Argentine laws cannot convert this country into the asylum for all criminals and degenerates of the world. It is fair and just in the defense of our civilization, that we should be inspired in the example of the United States. . . . It is absolutely necessary to resolve without vacillation the problem of social order and security, raised by the actions and propaganda of those who under the sympathetic flag of labor introduce themselves into our hospitable country to obstruct the functioning of the institutions upon which our free and civilized national life rests.[50]

Juan Alsina, in his thesis on immigrant hygiene, had included character, racial traits, age, and condition of immigrants as well, as criteria for admission. He pointed out that "with the goal of avoiding begging, the immigration law does not permit entrance to the country of individuals over the age of 60 or unable to work, whether because of deformity, chronic disease or dementia, unless the person has relatives who can guarantee sufficient means to provide for their subsistence. Otherwise, and we have already seen many examples, these individuals will come to be charges of the State, which should not support them."[51] Captains of ships could refuse to take on passengers over the age of sixty or those they determined to be "useless." Alsina, who associated poverty and dirt, noted that on many ships laundries were available but that they were used "disgracefully little" by the immigrants. The law, he said, prohibited the entry of dirty clothes or rags. Alsina linked hygiene to class: "As the majority of immigrants are very poor and uncultured people, they are unaware of the most elementary rules of hygiene, and they keep with them the filthiest scraps as if they were valuable garments. We in the Immigrants' Hotel are able to recognize the origin of those who arrive in this state of misery and filth, especially those who come from Brazil, from the condition of their clothing and persons."[52]

Two decades earlier, in 1876, the Argentine general inspector of immigration had stated that "those in first class are not considered immigrants." "Immigrants" were second- and third-class passengers; they alone were inspected and classified by class and occupation. First in priority were the agricultural worker, considered the "true immigrant, he who like the Nile fertilizes the country wherever he settles, grows, and multiples, arriving to be the founders of future nations." Second were businessmen, men of letters, and engineers, though presumably most of these individuals traveled in first class. Then there were the "unemployed from the big cities, compared with the first group they are tributary streams, without a doubt useful, but you can barely consider them immigrants," since this group was likely to return to Europe after trying to make a fortune in Argentina.[53] A 1898 report indicated that, in the early part of the nineteenth century, there had been an effort "by the government of Buenos Aires and President [Bernardino] Rivadavia, to bring the first immigrants, men of letters, professors, teachers of arts and industry, and families destined to occupy the ranches and agriculture."[54] By the late nineteenth century, however, those primarily needed were industrial workers and peons for the estates.

Since work skills and physical health were prerequisites for entry, and good work habits were considered necessary for future citizens, upon arrival immigrants were "meticulously interrogated and classified" according to their ability to work. Those who intended to labor in the rural countryside received visas allowing them to stay. Those without means or plans were taken by coach to the Immigrants' Hotel to be registered. There, men were separated from women and children, and each lodger received a coupon that provided five days of food and lodging in the hotel. As part of their stay, and "from the moment of arrival," adults were interviewed by the

National Work Office, which attempted to place them in jobs. Those who left the hotel without set employment seemed destined to "experience their own indolence and inability to work."[55]

The work office observed that sick immigrants were incapable of work and had diminished "industriousness and diligence"; such arrivals lingered at the hotel and it was "a struggle to set them loose," but they were allowed to remain in the city, despite the fact that they were believed to crowd the slums and bring more disease. In extreme cases, immigrants with chronic diseases "that disabled them for work" were "rejected and marked," that is, put back on ships to be returned to their port of departure. It was necessary to expel them because "if they were accepted they would be a threat" and cause "the degeneration of our race." An 1890 Immigration Department report observed that, thanks to the work of department physicians and the additional medical experts who were summoned as needed, "disagreeable circumstances" had declined, "due to the energetic means adopted by the Government in declaring invalid the contracts that are poorly fulfilled, sending us people in bad condition that we do not need."[56]

In 1875 the employment office at the Immigrants' Hotel had issued a report tracking the destination of newcomers by country of origin and their professional status. The next year, the commissioner general of immigration, describing attempts to regulate immigrants' arrivals, said that, following the United States model, a representative would board each ship along with the sanitary inspector and offer lodging in the Immigrants' Hotel to any passenger who needed it. Once an immigrant was in the hotel, dependent on the staff for food and lodging, it was relatively easy for inspectors to examine them. But, by 1890, officials reported that the hotel was holding as many 5,000 individuals in a facility built for 2,000. Difficult to ventilate, the wooden building was also difficult to disinfect, and flammable besides. A special building was needed for patients in isolation.[57]

After 1880, whether interviewed and surveilled at the Immigrants' Hotel or not, admitted foreigners were increasingly blamed for burgeoning problems of crime, urban crowding, and social unrest, including connections to anarchism and other forms of social disruption. Criminologist and national legislator Lucas Ayarragaray identified anarchism as "an exotic phenomenon in our country; it is an import," and explained, "If we let enter into the country a stigmatized population (since immigration brings in its productive currents the best elements of Europe, it also brings us the muck expelled by the old societies) . . . it is necessary to write legislation with an extensive and scientific concept in order to sanitize this population."[58]

In 1902 officials for the first time debated placing systematic restrictions on immigration to filter for dangerous types. On the floor of the national legislature, Deputy Mariano de Vedia proclaimed that Argentina defined immigrants as those "men of the world" who "want to inhabit our land" but said they "cannot be understood as those who bring us their vices, their resentments, their illness, the residue of European ferment, from the very nations from which they were deported and who arrive here in search of a new setting." Legislator Manuel Carlés explicitly identified crime and anarchism with the foreigner—"a foreign, bastard, shameful, and cruel mind that inspires crime! . . . It is not that which comes with its arms to enrich the land, that comes with its intelligence to teach science, that comes with its healthy example to mix its effort with ours to glorify the nation!"[59] Applicants who fell into one of a long list of categories were now to be denied disembarkation, including "idiots, mentally ill, epileptics, beggars," and those who for "whatever physical inability lack aptitude to provide for their own needs." Disease was construed broadly to embrace criminals, polygamists, prostitutes, "persons who aim to introduce or who exercise immoral business," anarchists, or professed assassins and bombers, the last barred from future Argentine citizenship.

In 1904 President Julio A. Roca proposed that moral requirements apply not to immigrants alone but to visitors to the country as well, and to anyone who "carries out immoral trades." A 1909 proposal cast a broader net to cover anyone who jeopardized "the security and liberty of work and with that the nation's economic order," with the goal of "safeguard[ing] the tranquility and prosperity of our working population and the progressive development of industry and commerce." J. Figeroa Alcorta and Marco Avellaneda, national legislators, recommended that the legislature approve a modification of the 1876 immigration law to prohibit "the entrance of immigrants with contagious diseases, unable to work, the demented, beggars, criminals, and those over the age of 70." While in their view this proposal did target second- and third-class ship passengers in order to "defend the country against the possible incorporation of pernicious elements," Alcorta and Avellaneda feared that the incoming tide was too strong: "Experience of more than a quarter century demonstrates that this legal precaution is not enough to impede the numerous individuals of poor conduct and ineptitude who move from foreign countries to our own, to the notorious detriment of our culture and social order."[60]

So, too, in 1909, Deputy Cayetano Carbonell insisted that immigrants be separated by positive and by "dangerous" traits. The existing law was too weak: "Today, our physiognomy as a cosmopolitan, ethnic, or demographic entity has changed," he argued, citing "the healthy and farsighted instincts of our statistics" as evidence for new legislation that would ensure "the order of public peace." He called on public officials, consuls, and immigration agents to be particularly vigilant in "surveillance, selection, and police order" to carry out "this great work of health, prosperity, and national defense." He exclaimed, "We cannot, as we do today, march ahead [while] unconscionably paying for the parasitic poverty or criminality of an aged Europe . . . . The only explanation is our deep-seated administrative negligence." He concluded that the immigration situation was "shameful to the national spirit, the sad and depressing spectacle that the republic offers us, invaded by imported legions of octogenarians, vagrants, beggars, criminals or suspected anarchists," who filled the jails and asylums, those "fertile incubators for crime at its peak and for growing and protected indigence."[61]

In his 1912 book, *Socialismo argentino* (Argentine socialism), Lucas Ayarragaray proposed as the best solution to Argentina's labor and social problem "selective immigration with scientific criteria."[62] Ayarragaray lamented the supposedly chaotic manner in which "residues of old and extenuated races" entered the ports and mingled with the indigenous population, "creating truly deplorable ethnic conditions." He proposed an authoritarian remedy of controlled immigration. He envisioned for Argentina a population of "good origins, having excellent types for progenitors, if not from homogenous races, at least from indigenous, mestizo or European ones without stigmas."[63] The following year Carlos Gómez, another national deputy, combined the idea of Latinness [*latinidad*] with Eurocentric views of Argentina's ideal national character. In the future, he said, Argentina had the potential to be the largest country of "Latin origins, like the United States is the most populous country of those who speak a language of Teutonic origin." Later he wrote that an Argentine traveling abroad cannot help but note the

> enormous superiority of the climate and fertile land of his country above all others he observes. She is undeniably destined to be the greatest of the associations of Latin origin. All the races, civilizations, ideas, beliefs are melding in this immense melting pot in our land, and . . . a rising national type is improving every day, as a physical and moral type. The Argentine race of the future, which is forming to lead civilization in this part of the world, will be Saxon in its audacity and its tenacity in great enterprise, and will conserve from the Latin the

creative faculties and love of classical beauty. . . . One Yankee statistician, after studying us, exclaimed that we the Argentines are at the
same time Latins and Saxons.[64]

Patriotism, Hispanic identity, and moral meanings of national identity
emerged as the true measures of Argentina's social pathologists. Their science was seemingly secular, but it revealed the influence of old ideas about
social order and access to national wealth. As scientists, they first checked
for diseases, but they brought with them deeply embedded broader political, economic, and even religious agendas. Their medical inspections,
which instructed immigrants both explicitly and implicitly about how to
be hygienic, clean, and free of disease, morphed into moral and political prescriptions and proscriptions. Along with medical advice they dis-
pensed high-minded ideals, expectations, and instructions on how to be
good neighbors, good parents, and good workers. The political economy
and the future of Argentina rested, in their view, on the unity of the national
body, racially and spiritually.

Physicians' and reformers' ability to help the nation's vulnerable immigrants was restricted by their client relationship to the state. As they
checked for diseases and racial stigma, the implication for healthy and ill
alike was that the state was monitoring their bodies and their behavior.
Psychiatrist José Ingenieros seemed aware of this when he suggested in his
book *Criminology* that immigrant "prophylaxis" was a vital component of
any program of "social defense."[65] Inspectors conveyed labor and political
ideals, and hidden not too far beneath the surface were moral and legal
messages about who had the right to belong to the nation. The emphasis on
health, cleanliness, and industriousness was shot through with traditional
ideas about the urgency of preserving the social order on which the political
economy and the future of Argentina rested.

Fingerprinting Foreigners to Inoculate against "Pernicious Elements"

As in the attempts to control the criminal population, the new method of
fingerprint identification developed by Juan Vucetich of the Buenos Aires
provincial police in the 1890s was a promising tool to monitor immigrants
with dangerous backgrounds. Called "dactyloscopy" in Argentina, the procedure swept through state bureaucracies and was rapidly absorbed into
criminal justice structures. While its proponents endorsed new fingerprint

techniques as "egalitarian," they in fact suggested its use first on groups deemed inferior or threatening. Immigrants were singled out for identification as potential future dangers to the state, the first noncriminal group subjected to fingerprinting. They served as a prototype for eventual universal identification projects. Fingerprinting of immigrants promised to isolate potentially dangerous individuals from the mass and expose any false identities.

Vucetich and his student Luis Reyna Almandos, also a prolific writer on the social application of dactyloscopy, led the campaign for mandatory fingerprinting and deportation of immigrants. Reyna Almandos wrote that "the influx of foreigners to our land results in grave problems. Welcoming them as one of the greatest benefits, we feel nonetheless the need to create laws and establish regulations in order to impede the entrance of pernicious elements that raise the crime rate." He suggested establishing offices of identification in the nation's ports of entry, where incoming persons' fingerprints could be cross-checked with international records of those with criminal pasts. "Completing this identification with those of the immigrants, expulsion could never be faked."[66] Further, it was assumed that fingerprinting could help track individuals with an inborn tendency to crime. The legal experts who forwarded a proposed revision to the criminal code of 1891 reflected a common, though not universal, view that the most dangerous crimes were committed by foreigners: "Our country is populated by European immigrants, and all foreigners know that in their country the penalties are hard and strong and work effectively. Therefore, we have determined in many cases to raise the maximum penalties, and adopt life sentences. . . . The biggest criminals in recent times who have moved the public attention were foreigners. Europe, in spilling over her excess population into our country, enriches us with its civilization and industry, but also infects us with its vices and evils, through an inevitable law of all immigrations."[67]

In 1912 the national immigration service established the Registro Dactiloscópico del Inmigrado, the Dactyloscopic Registry for Immigrants, as a repository for immigrants' fingerprint records.[68] They ordered officers at the Office of Work and Dispatch at the Immigrants' Hotel to complete a form, titled "Description of the Immigrant," with anthropometric description and fingerprints, for every person entering Argentina.[69] Each form corresponded to a number on the "immigrant's book" issued to each immigrant, with space for fingerprints and a photograph. The book also included information about rights and responsibilities, education, national currency and banks, and the location of steamship companies. As long as the immigrant carried the book, that person was promised protection under the law. In tones both welcoming and warning, the pamphlet said, "The honest and hardworking immigrant is a welcome guest. . . . He is an efficient collaborator in our ceaseless and marvelous progress, a healthy element that adds to the national family with its aspirations to economic improvement. This interest results in assimilation and permanent rootedness."[70]

Some argued that public crimes, linked to crowds and immigration, were "contagious," leading to social chaos, and should not be protected by the national constitution, because the constitution allowed group association only with "beneficial goals [fines útiles]." Thus, the text of the 1891 draft of the criminal code stated, "He who publicly instigates to commit a punishable act, while not assuming the character of complicit agent of an attempted or consummated crime, commits a grave act of repression, because he produces a true danger to the general peace, exciting the passions or criminal tendencies in an indeterminate number of people."[71] A 1913 proposed section of the criminal code went further, suggesting preventative detention, especially of foreign perpetrators, as "an imperative requirement of public order." All countries faced the dilemma of the tension between the need for order and the right to liberty, but such a predicament was especially acute for Argentina: "The difficulties increase in young and extensive countries such as ours, inhabited by a cosmopolitan population, migratory and nomadic . . . a country in which the urban centers as seats of legal authority are spread out over enormous distances and where the police, even at the lowest levels, are required to obtain permission in order to contain criminality."[72] Francisco Durá, a leading legal scholar, suggested a modification of the immigration law (which had not been fundamentally changed, he pointed out, since 1869) that would require "the most perfect personal identification of the immigrant." The key to this identification would be fingerprints. To aid "social defense" efforts, one copy of the new arrival's fingerprints would be sent to the immigration office, and another to the chief of police of the location to which the immigrant was destined.[73] Such measures would be preceded by extreme and unprecedented legislation against foreign radicals that rested on the widespread theory of, in Miguel Cané's words, "contamination" by immigrants.[74]

Like the immigrant medical examination, the fingerprinting of foreigners was largely symbolic and disciplinary, and of limited effectiveness in practice. The scope of state actions was limited by state incapacity to intervene in the great mass of citizenry. There were just too many people in the city, and it was inefficient and impossible to surveil them all. An under-funded criminal justice infrastructure made the procedure useful to police largely as a scare tactic, reminding newcomers of their subordination to the state and of the expectations for their behavior.

NINE

To "Formulate a New Race, the Argentine Race," for Democracy and Civic Regeneration

> I am sure that if we incorporate the foreign element to Argentine nationality
> with decisiveness, loyalty, and confidence, we will have created the bases to lay
> down in solid form order, peace, and liberty in this country, and we will have
> closed the era of inorganic politics.
> —LUCAS AYARRAGARAY

In the 1890s the Argentine newspaper *Tribuna* published a set of essays by
distinguished judge and legislator Agustín Alvarez. Appearing as a series
called "Manual of Argentine Imbecilities," the essays were a scathing critique of what Alvarez saw as the chronic instability and uncertainty in
Argentine politics. Alvarez was one of Argentina's most active politicians;
he had devoted his life to public service in the military, law, and as an
elected official, and like many of his colleagues in politics, he had an interest
in the intersection of medicine and law. He saw social problems in medical
terms, and he applied the newest psychological interpretations of behavior
to the local political scene. Publishing the essays in book form in 1899 as the
Manual de patología política (Manual of political pathology), Alvarez introduced a forensic political study of the Argentine citizenry and its potential
for civic responsibility. To Alvarez, "the thing that is most interesting and
worthy of study in this country is not the River Plate, nor the pampa, nor
the mountains, flora, or fauna . . . but the Argentine citizen, the type of man
who has produced on our earth 'Argentine ideals.' " Drawing an analogy to
horse breeding, he wrote of the Argentine people as a "race in training" for
democracy and civic action, which were necessary to build the foundation
for modern politics.[1]

While admitting large numbers of immigrants, the Argentine state fought
a rising tide of pressure to include them and broaden their participation in

civic affairs. Most elites believed the masses were incapable of good political decisions. Elites, claiming to desire democracy, did, however, support the opening of elections and sought other ways to assimilate or transform immigrants into citizens with a sense of love for Argentina and their civic duty. To do so, elites invented a new language based on racial imagery from the nation's past. They tried to rewrite the nation's education goals and legal codes according to what they considered the most modern criteria. The laws that resulted, however, were as exclusionary, if not more, than before. State modernizers limited civic participation of groups whom they considered inferior. Immigrants welcomed into the country as laborers were nonetheless prohibited from voting or holding office. After the Sáenz Peña Laws of 1912, which allowed secret balloting for male citizens, efforts to control and limit the participation of women, persons of both sexes considered criminal by nature, and all immigrants persisted well into the mid-twentieth century. But legislators, citing the studies and proposals of their physician colleagues, went further than denial of citizenship; they also put through laws to segregate, subordinate, and politically erase undesirable social groups.

The justification for exclusion of some from citizenship rested on scientific theories of who was fully rational and who was fully capable of loyalty to the state. State officials called on scientists to define the boundaries of citizenship and to provide the theories and language of inclusion and exclusion for rewriting the nation's laws and policies. Scientific expertise was used to determine who was rational, who could participate fully in politics, and who could belong to the nation as an Argentine citizen. The political ruling class, while denying rights to the majority of the population—all women, plus many immigrants—saw an opportunity to extend citizenship to the masses of malleable children of foreigners who might become a new, loyal, and true (male) citizenry in the second generation. Naturalization and the inculcation of patriotism and Argentine values were two broad strategies. Through new legislation regulating requirements for public education in 1884 and obligatory military service in 1901, officials specifically hoped to nurture a citizenry in conformity to national ideals.

Modernizing the nation's legal codes included attempts to control who could become a citizen and how, as well as the legal reform of codes that focused on the regulation of individual behavior. Legal modernizers also sought to maintain and rigidify a gender hierarchy through the law. Traditional views of individual responsibility existed in a tension with a determinism that claimed that some individuals and groups had no control over their actions but were in the grips of pathological states that determined their behavior.

Because elites believed that women and most men were incapable of rational political participation, they denied full citizenship and civic responsibility to all but a tiny percentage of the population. Women could not vote until 1947. The vast majority of male immigrants, by 1900 at least a third of the population of Buenos Aires, in addition to large numbers in other cities and provinces, had no formal political rights at all, and after 1902 they were vulnerable to surveillance and even deportation for subversive political activity. An additional group—the poor, mentally ill, and sexually deviant—found their freedoms curtailed as courts increasingly judged them incapable of rational choices. Experts argued that incapacity was rooted in a range of biological and psychological anomalies, which, thanks to the advances of modern medicine, could be scientifically measured and verified. The civil and legal status of all individuals was thus parsed according to a variety of biological, social, and political criteria. Immigrant men were made to wait for citizenship, with full participation denied until their sons had grown and could exercise it for them. Nor was the question only who could enter the nation but who could become permanently a full citizen. Who was a "real" Argentine? Men who held an Argentine passport could vote and serve in the military. Women could do neither. All women were denied the privileges and responsibilities of citizenship.

Weighing and Measuring the Words of Law: Legal Codes and Civic Responsibility

The drafting and redrafting of the national criminal code at the turn of the nineteenth century, ordinarily an academic and political routine, electrified the country to debate responsibility and social danger with the urgency of national purpose. The criminal code was a central instrument to establish who could act in the legal sphere and in the public sphere. Any new version of the codes in Argentina, as in all of Latin America, was based on a system typically separated into criminal, civil, judicial, commercial, and labor codes. Modifications had to be ratified by both houses of the national legislature and signed by the chief executive before they could become law. A signal goal of the Generation of 1880 and its allies for legal reform was to alter scientifically and, thereby, improve the law, as if it were medicine. Legal expert Francisco Durá, speaking of the importance of legal instruments to affect social relations, stated in 1911, "The words of the law, especially those of penal law, have to be measured and weighted with instruments of the greatest precision, like the milligrams of toxic substances that are used in pharmacies for the medication of men."[2]

The defining of criminal behavior and its punishments, and of codifying criminal law, which in turn led to definitions of who could be a citizen and true Argentine, began with the national reorganization of the mid-nineteenth century.[3] The Constitution of 1853 had authorized a penal code; eleven years later Carlos Tejedor, the leading legal scholar in Argentina at the time, was appointed to draft it. The new code was finally sanctioned in 1887, but other legal experts immediately began to campaign for amendments on the grounds that Tejedor had produced a code hopelessly out of touch with modern theory. They had hoped to eliminate all vestiges of colonial law, but the draft retained, in the view of reformers, the same old philosophical underpinnings. In the next thirty-five years, no fewer than six rewrites were produced, the results of contingent efforts by criminologists, medicolegal experts, lawyers, and progressive legislators, to bring Argentine law up to a modern standard. Each draft was published and widely read in elite political and scientific circles and debated in the legislature, often surviving multiple votes in Congress and committee without sanction.

Forward-looking jurists in the 1880s held that criminal justice must rest on a "scientific base."[4] The numerous *Proyectos de código penal*, or criminal codes that had been drafted over the years, were important vehicles for positivist and Social Darwinist ideology in articles by national deputies, senators, and government ministers that were published in scientific journals such as the *Archivos de Psiquiatría, Criminología, y Ciencias Afines* (Archives of psychiatry, criminology, and related sciences) and the *Revista de Criminología, Psiquiatría, y Medicina Legal* (Review of criminology, psychiatry, and legal medicine).[5] Lucas Ayarragaray, one of the most outspoken national deputies on the role of law in addressing crime and suppressing political ferment, published nine articles on the subject of the criminal code in the *Archivos* between 1902 and 1912. He adopted three core ideas directly from European criminology: the individualization of crime, the rationalization of punishment, and "social defense," or the need to protect society from harmful behavior from any quarter. The first two were necessary to ensure the third. Trained specialists could examine and classify criminal offenders, putting each one in his or her own criminal category, and then calculate the most effective form of treatment and formulate preventative measures. Science and medicine were considered the keys to this orderly process. As part of their broadly conceived program, criminologists lobbied for such judicial innovations as probation, jury trials, suspended sentences, and special measures in cases of reincidence (repeat offenses). One scholar trumpeted the impact of that approach when he wrote in 1904 that "the lessons of medical criminology have shaken the principles of penal philosophy from their base, and notably modified the corresponding positivist legislation."[6]

The scientific expert came to be understood as the linchpin of liberal democratic justice in Argentina and elsewhere in the Atlantic world. These elites were technocrats, who entrusted the real business of running the country to themselves and their carefully chosen experts. Psychiatrists and medicolegal specialists appeared as expert witnesses in trials, took an active role in defining the markers of criminal identity, and drafted new laws and policies to punish criminals and prevent future offenses. In a 1903 review in the *Archivos*, José Ingenieros wrote, "The purely legal study of crime is useless; physicians, pedagogues, etc., should exercise judicial function. . . . Judicial authorities will study crime as a social phenomenon, the criminal as a psycho-physical being." A key component in the rise of professional legal medicine was the growing importance of expert witness testimony in the courtroom. Francisco de Veyga's 1898 *De la prueba pericial y de los peritos* (On expert proof and expert witnesses) was commissioned by the Department of Hygiene. Veyga described how the judge determined the degree of responsibility for crime, but the legal medical expert offered evidence based on the new methods of criminal identification and conducted psychological and physical examinations of both the accused and the victims of trial cases. The scientific experts also performed autopsies, of course, but increasingly, especially after 1880, they also examined for physical evidence of crime, including fingerprints.[7]

A key task of medicolegists that had far-reaching consequences was to determine the mental status of defendants in criminal trials, a process that raised a number of profound, arguably unempirical questions about volition, responsibility, free will, and morality. Yet, positivism, a philosophy that emphasized scientific fact, verification, and objective truth, had changed the question of responsibility and free will dramatically over the course of the second half of the nineteenth century. In contrast to the "classical" school, most scientists now doubted the existence of pure free will and, consequently, ideas of guilt and of retributive punishment.

Criminal anthropology, with its deterministic explanations, further pathologized crime in the legal sphere. As medical doctors determined the defendant's mental state and degree of responsibility, they also intro-

duced ideas of social defense and dangerousness [*estado peligroso*] in criminal trials. In the context of the pathologization of crime, an important question for medical doctors was the difference between madness and crime,
and in the process, experts created the concept of criminal insanity [*locura
moral*].[8] It was of seemingly great importance to be able to distinguish
among the truly mad, who was unaware of his or her actions, the criminal
with "intent," and worst of all, those "simulating" or feigning mental illness. These differences were worked out in the courtroom and in the pages
of medical journals. Professional publications, and later criminological
ones, routinely reprinted medical reports from trial testimony for general
reading by physicians. For example, in an 1898 article in *Semana Médica*,
neurologist José María Ramos Mejía commented on the case of "G.E." and
the determination of his mental state. After a lengthy introduction on the
psychological and diagnostic methods used in his examination of the defendant, Ramos Mejía concluded that G.E. was not mentally ill and therefore
responsible for his crimes: "[The defendant's] will does not present percep-
tible change and he knows how to fake madness. . . . Even in his most
overexcited moments, G.E. retains control of his will; one observes not even
the remotest elements characteristic of melancholic rapture [*raptus melan-
colico*] that disturb the will of the lunatic and throw him into a series of
dangerous violent [acts]."[9]

Similarly, doctors considered the act of suicide a social problem, one that
by its nature blurred the boundary between medical and moral implications. In 1884 one expert called for legal action in the prevention of suicide:
"Considering suicide from the medicolegal point of view, we should rejoice
that modern legislation has erased from the Codes the monstrosities that
established punishment for suicide . . . . [T]here are two forms of voluntary
death: one that permits liberty or will intact, and another that attests to the
destruction of the faculties. The judge, the moralist, and the medical doctor
should consider the question of suicide taking into account these two points
of view, and define with just and firm impartiality those which correspond
to each one."[10]

The contributions of physicians, medicolegal experts, and criminologists, such as the famous criminal psychiatrist José Ingenieros and hygienist
and forensic specialist Francisco de Veyga, were central to code reform.
They provided the basic concepts of human behavior that informed legal
language. These authorities implemented reform as expert witnesses, asylum administrators, and scientific theorists and researchers. Social pathologists like Ingenieros had a vital role in the reform of a broad range of legal

institutions and in the founding of asylums for the criminally insane, for juvenile delinquents, and for released prisoners. Ingenieros's comprehensive plan of 1911 for the nation's "social defense," for example, included the modernization of criminal identification; judicial innovations of probation, suspended sentences, and special measures for cases of reincidence; new modes of prison discipline and quarantine; and "prophylactic" laws aimed at immigrants and the poor. In the *Archivos* Ingenieros, as editor, gave extensive attention to criminal probation, trial by jury, and the various Argentine legal *Proyectos*, or draft codes, and he regularly reviewed criminal procedures of other countries, as well as their treatment of beggars and vagrants, alcoholics, the mentally ill, and certain aspects of civil law, especially divorce.[11]

At the core of all scientific criminal code reform was the controversial revision of the concept of criminal responsibility. As in all Western criminal justice systems, the idea of responsibility was the linchpin of Argentine criminal law, and changes to that idea held up approval of new codes for years. Reformers aligned with the positivist and scientific movements attempted to reshape the idea of responsibility, introducing a more complex concept, imputability, a new way of measuring whether an individual should be charged and tried. The focus shifted thereby from classical notions of free will to, in their view, more scientific and measurable criteria of degrees of responsibility, linked to dangerousness. Late 1880s liberal criminologists were well aware that most jurists accepted the term "criminal will" as the "free decision to commit an illicit crime." The 1887 code reflected this in its Article 6, which stated that "in the execution of acts classified as crimes, we presume criminal will [*voluntad criminal*]."[12] Would-be code reformers proposed a thorough rethinking of the concept of responsibility as it stood in the code. As early as 1891, in an extensive commentary, well-known positivists Norberto Piñero, Rodolfo Rivarola, and José Matienzo offered a modification to the old Article 6, replacing "acts classified as crimes" with "punishable acts." This modification seemed to reflect the change from a moralizing to state-directed judgment. The need for scientific facts was stressed in the reformulation of the law because, in the authors' view, "it is necessary that the presumption [of criminal intent] rests on positive data . . . that is to say, from its skilled verification."[13]

In June 1890 President Miguel Juarez Celman appointed the same men, criminologists and legal experts Piñero, Rivarola, and Matienzo, to draft a new version of the code. Their introduction to the draft code began ominously: the existing law of the land "suffers defects that must disappear because of the danger that they pose to society." Positivists all, they tackled the age-old problems of free will, intent, punishment, and conditional liberty. Where no settled scientific opinion existed, the authors reassured the legislators that answers were not far off. For example, on legislation on personal responsibility they reported that "criminalists [*criminalistas*] continue to investigate the solution [to the problem of responsibility] in order to offer legislators a solid base in this very delicate and important matter." They included new articles on crimes against public security, public order, and national security, among them arson, bombings, attacks on railroads and telegraphs, and crimes against "public health," the sale of poisons or adulterated food and medicine, and sabotage to public works and hospitals that "generated a common danger of much concern." Their emphasis on the magnitude of crimes against public property, especially on means of transportation and communication, reflected the state's interest in protecting foreign capital investments and the burgeoning export economy.[14]

The 1906 draft code replaced any implication of "responsibility" with "intent."[15] Yet, according to its critics, one of the greatest flaws of the 1906 *Proyecto* was its failure to resolve the issue of responsibility and provide a formula for determination of intent. These critics wanted a more specific list of physical and psychological states that would give judges discretion to eliminate imputability. They called for judicial determination of degrees of responsibility based on individual traits and abilities. One commentator asked rhetorically in his review of the 1906 draft code, "How can we impose universal criteria on the immense variety of individual acts that have their origin in the profundities of consciousness?" He concluded that because of all the different reasons that people commit crimes, "it is totally impossible to establish a general rule that serves as the base of juridical presumptions. In penal law, one cannot isolate the will. . . . [O]n the contrary, one should study it in its psychological context and in its totality, to know its relation to the action of the agent."[16]

Determining Dangerousness to Ensure Maximum Social Security

The ability to determine responsibility rested on the routine physical and psychological examinations of prisoners, a procedure that had been urged for years and officially was recommended at the 1914 Argentine Penitentiary Congress in Buenos Aires.[17] Biopsychological examinations, conducted in prisons and police stations, and their extensive documentation were intended to serve not just as research tools but as objective evidence in court cases. José Peco, who later served as rector of the University of La Plata, expanded these new criteria to the entire criminal and potentially criminal population, advocating mandatory medical, physical, and psychiatric examinations of all who entered the criminal justice system. Factors such as age, level of education, "determining motives," background, and social factors were to be taken into account in determination of the offender's level of "dangerousness."[18] In 1919 a Senate commission appointed to review the draft criminal code noted that, above all, *"defensista"* theories (that is, those that subscribed to social defense theory) were concerned with the dangerousness of criminals and with formulating a calculus of, as the commission put it, "maximum possible social security with the minimum possible ill treatment of the individual." The very choice of the word "imputability" in place of "responsibility" fed the quest for "social defense," for the former meant "one's capacity for social behavior. Only when one has this capability, can one be charged as culpable with antisocial behavior."[19]

In 1921, when the code was finally sanctioned, legislators agreed that offenders should be subjected to psychological and medicolegal evaluation of their degree of dangerousness. From then on, offenders and perpetrators were to be judged not on their character or will but on their level of scientifically determined threat to society. The determination of responsibility was to be based on physical examination of suspects' minds and bodies, carried out by physicians and psychiatrists. Professional criminologists' emphasis on science, with new approaches to criminal identification and classification, and above all with their compelling idea that the nation must put its own interests above the liberty and well-being of individual citizens, established the criminologists' indispensability and assured them a unique role in the disciplinary offices of the state. The official acceptance of these ideas, in turn, paved the way for an expansion of classification in the criminal justice system and led to ever-finer categorizations of expectations for individuals beyond the original criminal subjects.

The most notable outcomes by 1921 were the by now uncontroversial acceptance of the idea that judges should consider all offenders according to their degree of "dangerousness" and the new reliance on "social defense" as criteria for determining treatment or punishment; the new terminology and the theories and techniques were now encoded in the law. The question was no longer whether legal subjects were responsible for their actions, or capable of them, but what their supposed level of "danger" to society was. Repression, an observer of a 1913 draft code asserted, was "not just punishment . . . but a means of social defense."[20] Adhering to the basic principles of the positivist school, authors of the code bragged that "the orientation of criminal law in our era is purely scientific and tends to deepen the study of the criminal as much as the crime itself." Referring to ideas introduced at the 1885 Criminal Anthropology Congress in Rome, the code proposed a number of innovations, such as preventative detention—"an imperative need of public order"—suspended sentences [*condena condicional*], individual punishment, and a new concept of recidivism, or repeat offense. The authors concluded, "All the new [categories of] crime that we created in this part of the Code respond to needs manifested in social order, and to the guarantee of respectable and primordial laws."[21]

Not all individuals were subject to the same criteria in the determination of legal status, however. Juan Ramos, the legal scholar, identified criminal alcoholics, the mentally ill, children, and women as having limited, special, or ambivalent status, each group requiring its own sort of treatments and designated institutions of correction.[22] (Nonetheless, researchers did not conduct a serious scientific study of female criminality until 1933.)[23] Officials, relying on the criteria of "dangerousness" and "social defense" as measures of offenders' status, thus suggested gender-specific treatments in the law and in other institutions of criminal justice, the civil code, property, marriage, and *patria potestad* or control of children; in general, these changes institutionalized women's dependent legal status. The debates over the criminal code revealed the prevalent assumptions about woman's nature and expectations of her abilities and determined how she would be treated in court and in institutions of confinement. The modernization and reform of practices in other aspects of criminal justice did not result in improved treatment of women. Prison modernization, rationalized courtroom procedure, and the regulation of brothels, for example, led to traditional and rigid gender assumptions about citizenship and the expectations of responsibility.[24]

On the surface, code reforms seemed to posit a gender-neutral criminal subject (except in cases of abortion or infanticide, when they generally assumed a female perpetrator). In reality, however, the criminal code was highly gendered. Most scientists and legislators believed that there were different levels of dangerousness for men and women, both qualitatively and quantitatively. No one—feminists and scientists alike—considered women dangerous in the same way as men, and consequently, they studied women scientifically in a very different way. Moreover, the maleness of the criminal was a central component of criminologists' theories, despite the proven ability of women to commit crimes, and women's real or potential criminal agency tended to be ignored. Instead, Argentine criminologists followed the Italian school line, focusing on women as mere "auxiliaries" or looking at a few defined "female crimes" (or quasi crimes, such as prostitution). They were more likely to consider women's biology and psychology as potentially pathological. Criminologists' enormous interest in hysteria as a female psychosomatic phenomenon was the most visible illustration of this tendency (see chapter 4).[25]

Psychiatrists, legal scholars, and criminologists who were hired to study female mental and reproductive pathology were key to judicial efforts to determine degrees of criminal responsibility and social dangerousness. Insofar as these experts assumed biological and psychological sex differences, they, too, held men and women to different standards of responsibility. Women's legal responsibility was confined to their capacity to raise productive, law-abiding sons and to maintain a moral home. The assignment of women to the home precluded their participation in the public sphere. While men accused or convicted of crimes were certainly denied their civil freedoms, women were denied civil freedoms at all times, a reflection of common beliefs about their diminished capability, as well as a scientific attempt to rationalize female crime and women's lack of legal responsibility. Scientists believed that, as a result of special reproductive roles, women were less responsible and less capable—if not incapable—of fulfilling important public roles. A common assumption was a weak will. Women's diminished ability to reason led to an inability to exercise free will in the same way as men did, thus the further conclusion that women were unusually suggestible and unable to reason normally. Scientists believed they could prove that women were in a "special" category along with children, the mentally ill, and (sometimes) alcoholics.

In general, psychiatrists and medicolegists denied women agency as criminals and, in doing so, denied them civic agency as well. The 1910 psychological examinations of two mentally incompetent subjects—one male, one female—illustrates this double standard. Examiners found that the man, an alcoholic with signs of hereditary degeneration, had "limited and perturbed" volition. They found the woman's will "profoundly altered." The male subject "is not his own master, is not free"; the female "sits without moving, without initiative."[26] They also cited the woman's sexual and reproductive history as evidence. A medical report wrote of "Señora A." that "her personal history is more illustrative. It appears that she had an agitated adolescence . . . but, she married and had no more disturbances. Then, despite her nervous temperament . . . her life transcurred without psychic abnormalities. But, at the age of fifty-two, perhaps by coincidence or by immediate effect, upon experiencing menopause she suffered an attack of mental disease."[27] It was assumed that such conditions had a lack of will in common, or in Ingenieros's words, they were "illnesses of volition." Ingenieros, in his studies of hysteria and hypnosis, cited the "instability" of the hysteric and "the importance of suggestion as a cause of [hysterical fits]."[28] In reducing women to their sexual and reproductive identity, scientists essentially removed them from the acknowledged spheres of rationality and public action.

At the turn of the century, for all such reasons, Argentina's women were entirely disfranchised. But, as the new century dawned, intellectuals, politicians, and activists began to debate women's potential for public roles in the context of the arrival of "modern" political and social structures both at home and abroad. The status to be accorded to women in law was a key component of this discussion, for it—at least in theory—provided a national template against which judges, legislators, and scientists could measure women's behavior. Despite a new awareness of feminism and woman suffrage movements in the last years of the century, however, gender roles in Argentina remained rigid. Women were expected to bear children, feed them, and care for their husbands. Typical for a patriarchal society, men were the official heads of the family, the sole breadwinners. They (in theory) made all decisions at home as well as in the public sphere. Men were responsible for participation in politics, for voting; women were excluded from both by law. Women's participation in the workforce was problematic; their political action was uncommon and suspect.[29]

Consistent with women's diminished participation in civil society was physicians' viewing their female patients in infantilized terms. Psychiatrist Bernardo Etchepare, in an article published in 1907, linked hysteria with "puerilism," or the state of being childlike. He offered a number of case studies of hysterical and alcoholic patients (all women) who had regressed to a childlike state. Etchepare referred to "the frequency of puerilism in women" and added for emphasis that "the observations published to date refer almost always to women."[30] Psychiatrist Francisco Netri concluded that "women live all their lives like *big children*, an intermediary species between children and men, which makes them singularly adept at nurturing and educating young children."[31]

The articulation of gender roles and expectations in the nation's legal codes revealed the tensions between tradition and modernity that any consideration of public behavior evoked. Women, like immigrant men, were not accepted as full members of the polity. While scientists had offered a novel approach to legal reform and introduced a new level of terminology, they held some remarkably traditional ideas about capability and participation that found their way into the nation's most fundamental legal codes. Responsibility, imputability, and agency appeared to be universal, but full competence was limited to men. Scientists and officials alike stripped women of their criminal agency, just as they deprived them of full civil participation. In the course of legal reform, culminating in the 1921 code, the core beliefs about responsibility had been transformed, dissolved into a more complex concept that now applied to all citizens. In this way, liberal legal reform, far from uplifting citizenship for the majority of the population, eroded it.

"Cover Them with the Flag": Naturalization and Citizenship

By 1912, middle-class reformers from the Radical Party, along with more revolutionary parties that represented workers and immigrants, argued that Argentina was ready to embrace democracy and expand political rights to a broader spectrum than the entrenched oligarchy. The daily newspaper *La Nación* agreed: "This nation used to be branded as indifferent, apathetic, and incapable; today, the nation responds to the patriotic call with the same generosity with which it used to pardon official corruption." A few days later, they noted that "the most rebellious skepticism has been conquered and the most persistent indifference dominated. No one resists the wave of civic-mindedness that has surged from certain partisan nuclei [and] extended its vibrations across the popular mass. The concept of free election, forgotten for so long, has come to impose itself with sharp clarity in all spirits, and each citizen is intimately aware of his patriotic duties."[32] The boundaries of citizenship were changing and reforming as previously disfranchised groups—women, noncitizen male workers, and immigrants— demanded a share of the country's prosperity and a voice in its governance.[33] While they had no vote, immigrants, women, and other groups made their voices heard through newspapers, ethnic mutual aid societies, demonstrations, and memberships in socialist and anarchist parties.[34]

In February 1912, after months of legislative debate, the Argentine government passed a series of sweeping electoral reforms laws named for President Roque Sáenz Peña. With the passage of these laws, the nation experienced for the first time secret, and mandatory, elections, with voter lists coming from military rosters. Political leaders believed that Argentina, while rife with social conflict, had progressed to the point where universal male suffrage was unavoidable. The Sáenz Peña Laws would create, in the words of the president, a state of "civic regeneration."[35] The president, a loyal son of the ruling oligarchic party, the Partido Autónomo Nacional, also sought to deflect the rising tide of challenge to his party's monopoly on national politics, in particular, the increasingly popular middle-class reformist and left-wing confrontations. Officials in power hoped the Sáenz Peña Laws would be a safety valve that would provide more people with a feeling of participation without truly threatening the political order. They correctly assumed that the middle-class Radical Party could be kept under their control.

Naturalization was key. It was the criterion for electoral voice. By limiting opportunities for naturalization, the elite could continue to control who could fully participate in electoral politics. Although there was a lively debate over female suffrage after 1900, it was not taken seriously by most. Moreover, by law Argentine women who married foreign-born men lost their citizenship; foreign women who married Argentines lived in a similar legal limbo. Any challenges to these practices served to underscore women's inability to claim full political rights and their status of "dependent citizenship." Ironically, the stripping of citizenship from foreign- and native-born women alike was justified with references to modern and liberal theories of individual free will. In an 1897 court case involving a woman who married a foreigner, the judges' opinion read, "When a woman marries a foreigner, she knows that, by her marriage, she becomes a foreigner, and she consents implicitly in the renunciation of her nationality and the acquisition of her husband's." The removal of the wife's citizenship, seen as her choice, was a reflection of the patriarchal understanding of marriage and women's status more broadly.[36]

The incorporation of foreigners into the body politic thus emerged as the central issue. The state's ability to select its citizens and to exclude dangerous elements at the border appeared to be an uphill battle in the face of the sheer number of immigrants entering the country. Even those who had championed mass immigration in the late nineteenth century were stunned by the overwhelming amount of newcomers in the first years of the twentieth. A tidal wave was poised to engulf their imagined national type. Many of the foreigners had growing ideas of their political rights and willingness to voice demands for better treatment by employers and the state. The ruling elites feared the potentially radical agenda of immigrant groups, but they also feared reform and the possibility that they would be forced to share, or even abdicate, power. Aware that their nation's progress depended on the flow of foreign labor, capital, and culture, many in the Argentine elite were ambivalent or even hostile toward foreigners, seeking to wall off the social and political hierarchy from interlopers.

By 1900, immigration had clearly exercised very different effects on Argentine society than the founding "liberals" had intended. Although, in light of the nation's ambitious export goals, immigrants were welcomed in Argentina as industrial laborers, as agricultural peons, and as bodies to fill the "empty" lands of the interior, it was another thing to welcome them as citizens, and this is where the real selection took place. Argentine state elites exercised exclusion less at the border than at the moment of naturalization. The oligarchic state, firmly in control, sought to maintain its monopoly on electoral and legislative politics, even as it proclaimed the nation a democracy. State-building elites and reformers believed that the average voter was incapable of electing the right parties to office, and thus they led their nation with the broadest, if only theoretical, guidance of the male electorate.

As city and state officials awoke in the early 1900s to the reality of uncontrollable mass immigration, they increasingly focused on shaping and disciplining those who had been admitted. Class-based conflict, on the rise, prompted the need, in their view, to segregate politically the existing population to control access to power. The most effective strategy of resistance was to withhold citizenship. In a series of residency measures passed by liberal governments after 1860, immigrants had gained business and property rights, but political naturalization required an arduous process. Police officials routinely advised applicants to abandon their requests.[37]

Government officials and members of the literary and scientific elite over the next several decades debated whether extending full citizenship to immigrant men might strengthen or weaken the nation. In a debate in 1890 over proposed amendments to the Citizenship Law, Deputy José Miguel Olmedo argued that the country should be concerned not merely with bringing in economic producers: "we ought to principally try to incorporate into our national life the man of thought and study, who brings to our land capital more valuable than all others: intelligence and activity destined to serve great and noble goals." In Olmedo's view, such immigrants were necessary to "continue the revolution that this country has completed in part."[38] Similarly, on the floor of Congress in 1898 Deputy Miguel Morel proposed making citizenship easier to attain. The most important thing was to assimilate as many foreigners as possible, he argued, and instill in them a love of country and sense of duty, military or otherwise:

> We have spent considerable sums of money to attract men from various parts of the world; we concern ourselves to give them all the possible things they need to live, work, and do well here, but we never concern ourselves enough to make this country a homogenous mass, a unified entity, a social and political entity able to resist all the commotions of its own organic transformation and at the same time to endure the vicissitudes of history. In a word, there is no progress, no civilization, no liberty, no country, if the foreigner does not convert into the Argentine. Not through violence, but through the virtue of the law of love, community, confraternity.[39]

An easier and more inclusive naturalization process, Morel concluded, would "cover [them] with the flag, declare them irrevocably its sons. . . . [I]n this way we will formulate a new race, the Argentine race: united, indivisible, eternal, and destined to realize in its time the most great and fertile transformations."[40]

Another supporter of immigration, Deputy Francisco Barroetaveña, argued that unnaturalized immigrants, estimated to be 25 percent of the total population, were a "serious ethnic, political, social, and economic problem." In 1909 Barroetaveña called Argentina "the second metropolis of the Latin race." "We have nothing to fear," he claimed, "from the ethnic, biological, social, artistic, and historical influence of the fusion of the mass of strangers rooted in our citizenship."[41] Immigrants must be assimilated as quickly as possible to unite the country. Adolfo Dickmann, a provincial deputy, spoke in 1915 on the floor of the Buenos Aires Province legislature in favor of reforming naturalization laws to include more foreign workers. A socialist, he wanted foreign workers to participate more in politics. "For our part," he said, "whether we are already Argentines by birth or naturalization, there exists a unanimous sentiment to contribute to the means of our power, to raise up Argentine nationality so that it figures in the concert of civilized nations." According to the legislative transcript, Dickmann's speech brought "prolonged applause" from the audience.[42]

But the majority in the legislature and political circles urged caution. For instance, Bolivian diplomat Santiago Vaca-Guzmán, observing the Argentine political scene, was opposed to allowing naturalized citizens to fill public office. "Is there a danger or not in the substitution of national legislators with foreign ones?" he asked; "Yes, there is, and serious." With Argentina overwhelmed by foreigners, the majority of elected seats would fall to foreigners. If the "adoptive citizens [who] would legislate and govern us were men of integrity, upright, inspired by all that leads to the greatness of the Republic," there would be no problem. However, "on the contrary, the straightforward precedents of democratic life lead us to presume that the government does not always fall in the best hands, but [rather] in political struggles the audacious and sly generally triumph, and only exceptionally the intelligent and meritorious."[43]

As the perception of foreign engulfment grew, nativism and xenophobia became more and more common, and the halls of the legislature rang with calls to bar from citizenship immigrants who displayed a range of undesirable traits. Deputy Lucas Ayarragaray proposed in 1908 that ten years' residence be required before naturalization, except for property owners and those who had married Argentine women and had children in the country. "When one has such an amount of foreigners as exists in our country, this composition of population causes problems of social order, economic order, political and historical order, that we should resolve with a law of naturalization," he argued. He would exclude undesirables from naturalization, especially "those foreigners who in or outside of the country were sentenced for crimes, or who had been expulsed from any nation as anarchists." Also problematic were immigrants who seemed to "consider the country from the exclusive point of view of their material interests." Naturalization should revolve around loftier goals: "Nowadays, the foreigner is a purely economic element among us, a species of living machine, an instrument of production who, if he has no civic love, it is because he does not feel himself a citizen." Long residence could mold the character of the immigrant:

> The bill I have introduced has as its fundamental criteria residence or domicile, because in the universal legislation of today it tends to be the main base determining civil law, and it appears to me that in countries of immigration like our own, it should be that which secures political rights. I have believed it necessary to establish a long residence on the part of the foreigner in order to achieve citizenship, because this act implies an express will to definitively take root, and because in this time period his spirit has surely molded to our peculiarities and the sap of our land circulates through his thoughts, ideas and generous inspirations.

The ten-year residence requirement would "constitute a sufficient guarantee and a defense for us."[44]

Similarly, national deputy Cayetano Carbonell recommended a probationary period of five years' residency or other requirements before citizenship. Preferred would be an affluent immigrant, one who either had remained long enough to make a good living and contribute to the economy, or whose wealth could be "verified," or who had contributed to the nation by marrying an Argentine woman and producing children "born in the country." Such a checklist would reduce the likelihood that "insignificant people" would obtain citizenship cards to register to vote and exchange them for money. The worst specter of liberal citizenship laws, in Carbonell's view, was the possibility that "more than one dangerous foreigner becomes naturalized with the goal of avoiding the legal action of the Residence Law and its severity."[45]

Increasingly, the elite's ambivalence toward foreigners, especially southern Europeans (from whom, ironically, most of them were descended), reflected competing views, both exclusionary and inclusionary, but with a common goal: a united, racially homogenous Argentine population infused with patriotism and devotion to the new republic and its economic progress. How could they create such a cohesive Argentine identity? They turned to the children of immigrants for the future of Argentina's citizenry.

The "Intelligent Incorporation" of the Immigrant

Juan Alsina, a physician and immigrant medical inspector, wrote in 1895 in the introduction to the second national census, "All these great attractions should bring about an enlargement of immigration and colonization of the Argentine fields. The Latin race, especially, will give some millions of its children to realize here a perfect and lasting *Christian Republic*, that to which philosophers and politicians have aspired."[46]

And yet, the principal problem with immigration, most social scientists, psychologists, hygienists, and state officials agreed, was cultural heterogeneity. The supposed disunity and racial heterogeneity were seen as pathological, as leading to social instability and fostering crime. José Gregorio Rossi, police commissioner and later chief, wrote in 1903 that in Buenos Aires "we add to the natural causes of increased criminality . . . the heterogeneity and instability of the population."[47] The need for assimilation of foreign and radical elements seemed more urgent in an apparent epidemic of crime. Eusebio Gómez, in *Mala vida en Buenos Aires* (Lowlife in Buenos

Aires), emphasized a seemingly high proportion of youthful offenders.[48] Victor Mercante, a top national educator, noted of one immigrant juvenile offender: "The Argentine social environment did not affect him . . . [g]iven the unmodified racial spirit that dominates in him and his lack of acceptance of the customs of the Argentine people."[49] Disunity and diversity were seen as promoting mental illness as well as crime. Victor Arreguine, a physician and researcher, wrote in the *Archivos* in 1905, "The progression of suicide is a result of the deficiencies and imperfections of civilization. . . . The lack of social homogeneity in Buenos Aires or any other city contributes to the rise of suicide in general and among immigrants in particular."[50]

The antidote to such social instability was assimilation. Historian Carlos Alfredo Becú articulated in 1903 the widely held belief among intellectuals that "man needs, in order to live in harmony with his society, to assimilate to it, to make himself similar. . . . The inassimilable, that is to say, he who deviates from the common type, is cast out from the community."[51] Criminologist Cornelio Moyano Gacitúa agreed: "[I]mmigrant countries inundated by the migratory wave need, like an organism, to be able to assimilate this glut [*superalimentación*]. They need to know how to receive the immigrant, select him, distribute him, allocate him; in a word, to use him."[52] Political elites spoke of assimilation as a social and biological process of adaptation that could transform motley races into what Alsina called "civilized" citizens. He stated hopefully: "The tight link of matrimonial alliance has established a community of national sentiment, interest, and purpose between Argentines and foreigners. Native and European families are now melding into one; the Spanish and Portuguese surnames of the families of the first century are uniting with the new Argentine surnames, of Italian, English, German, [and] French origin, and of other nationalities. We see Argentines appearing in the Government, Congress, Legislatures, Schools, Army, Navy, Clergy, and in all of Society."[53]

Cultural assimilation seemed to have parallels in nature. José Ingenieros's review of *Traité de biologie* (Treatise on biology) by Felix Le Dantec commented that "the essential act of life, in its complex manifestations as in its elemental ones, is assimilation."[54] Another author believed that "the general idea of adaptation in biology reduces to this: acquire the ability to live in a determined environment and conditions, which entails a modification of the organism that adapts itself—or modify the environment in which it must live. The application of these ideas to social life obliges us to verify that, of the procedures of adaptation, the most frequent is the first."[55] Cultural absorption, adjustment, and nationalization of the foreigner were natural and essential processes in preventing social disruption.

In legal scholar Carlos Gómez's view, the government must assume an active role in engineering naturalization and citizenship to shape national character and foster patriotism and unity. He wrote in 1913, "Our responsibility and our task are to forge and chisel the race of the future, nationalizing it every day, every hour." "We need to nationalize schools, character, men, tendencies and parties," he continued, to arrive at "a *Nation* with *citizens* and not a conglomeration of citizens with a *Nation*." The goal, said Gómez, was to "found a Nation with political unity, with unity of civilization, with racial unity. All this is nothing other than the Argentine nationality of the future." Moreover, "if we aspire to consolidate and intensify national sentiment, we should take care with its formation and its greatest developments, by means of a legislation that inspires to such high intentions."[56] The Argentine elites sought, in the words of national deputy Cayetano Carbonell, to orchestrate the "intelligent incorporation" of the "right" kind of immigrant. The goal was a nation "not just densely populated, but also an attractive, homogeneous, and cultured center civilized by races of the entire world, that with its virtues of intelligence, capital, and labor and human morality, will contribute to our respect and greatness, according to a future universal concept."[57]

Argentine leadership, inundated with masses of southern Europeans, began to focus on the need for racial unity but became more selective and more xenophobic over time. While never fully letting go of the Nordic Anglo-Saxon ideal, elite policymakers encouraged programs that would overcome the nation's dominant southern European traits. One such effort centered around the hundred-year anniversary of Argentina's liberation from Spanish rule, celebrated in May 1910. The government erected monuments, planned parades, and created new nationalistic school curricula for children. Twenty thousand schoolchildren dressed in their best clothes were led through the streets, parading under fancy installations of Argentine flags. With a budget of over 13 million pesos, a special commission planned a series of events, among them an international medical conference and expositions on the nation's hygiene, industry, agriculture, and art, that showcased Argentina's advances in science and medicine, a visible sign of the level of civilization achieved in the hundred years since independence.[58]

Experts hoped to mold the confusing mass of polyglot workers and potential citizens through education of various types. The centennial panorama was directed at native-born and foreign Argentines alike, communicating new ideas (or really, hopes) of what it meant to be an Argentine. Neighborhood social clubs were formed with names like "The Pampa," "The Creole," and "Sons of the Desert," which encouraged people to celebrate their Argentine identity through song, story, and dance.[59] The famous Nicaraguan poet Rubén Darío, who lived for some time in Buenos Aires, described his host country with awe and optimism in a 1911 essay, "Argentina":

> A people formed from Spanish fathers who inherited all the qualities and defects of the conquistadors, together with a collection of new ingredients, initiated its independence with epic deeds, suffered through consequent disturbances and the revolts of a state in its testing stages, endured the winds of pampa anarchy, and bled profusely in civil disputes. It found out about the weight of lead and iron tyrannies and revolted against them. Little by little, it began lighting up its own soul, the soul of the people, and learned the true difference between civilization and barbarism. Argentina takes good care of its schools and universities, spreads culture and progress, raises up the parliamentary system, and sees that its greatest riches lie in the heart of the country. It is preoccupied by economic matters that are vital. By elimination and by crossbreeding, it is beginning the formation of a brilliant race.

Darío praised Argentina's "concept of national purpose" and "sense of the country's future." In his view, "work and culture" were the keys to Argentina's success in incorporating large numbers of foreigners.[60]

Even more directly, the army presented a means of assimilating men of the largely foreign-born population. It would step in for weak state institutions by taking the sons of immigrants and making them Argentines. Foreigners were seen as isolated from the "real" Argentine culture, living in enclaves and refusing to learn Spanish. Within the army, there had already been problems: soldiers in the nineteenth-century wars had to form contingents of Italian or German speakers since many spoke no Spanish; boys failed to register for the draft because they (supposedly) did not understand the law; and foreign-born soldiers' loyalty to Argentina was in doubt. State leaders in the early twentieth century hoped the bonds formed in military service would help assimilate boys from disparate cultural backgrounds, creating a common culture. They aimed to fill the gap in public education by teaching reading and basic skills to recruits. Military service would also purposefully teach discipline, morality, and citizenship, and recruits would absorb a sense of order.[61] While some second-generation immigrants chose to enter the army as a way of social mobility, in the end most avoided military service. In fact, many immigrants did not pursue naturalization just to avoid it. The 1901 Richieri Law, which made military service obligatory, was designed specifically as a way to inculcate large numbers of immigrant men with Argentine values, but many more came into contact with the armed forces through conflict on the streets, as the army was increasingly called out to support the police in Buenos Aires and other major cities with large worker populations.

The most promising path to mass assimilation was through the children of immigrants, who were seen as docile and malleable. State officials considered public education the best means of inculcating Argentine values and nationalistic spirit, as it reached immigrant children at a young and impressionable age. Nationalists of the early twentieth century took one of the central tenets of late nineteenth-century liberalism, strong public education, and shaped it according to the new conceptions of national identity. They regulated the curriculum in primary schools, no doubt aware that most children would attend only one or two years. They enacted legislation banning all languages but Spanish from the classroom. In 1888 the government insisted on six class hours a week of Argentine history, three times the previous amount. A few years later, under physician José María Ramos Mejía's leadership, the National Education Council required all teachers of Argentine history, geography, and civics (even at private schools) to be citizens. Ramos Mejía ordered school textbooks revised to emphasize Argentine heroism. In 1909 children were required to recite a pledge of allegiance, or "Oath to the Flag." The first duty "of a good citizen," the oath confirmed, was to "love his country." As in a catechism, the child was then questioned, "Even before his parents?" The prescribed answer was "Before all."[62]

Generation of 1880 members, among them Ricardo Rojas, a prominent intellectual, and José María Ramos Mejía, one of his generation's most prominent hygienists, led the early twentieth-century movement for education reform with a nationalistic agenda. Former president Domingo Sarmiento had famously initiated a system for modern education in Argentina in the 1870s, advocating the adoption of the U.S. model. Early twentieth-century reformers, by contrast, emphasized Spanish language and heritage, national history, and the studying of an invented national "culture," including the romanticized gaucho, an icon once reviled but now esteemed. National identity, they believed, must be communicated to the largely foreign population through literature and education. The earliest articulation of the concept of "*argentinidad*," or Argentine identity, appeared in the work of Rojas.[63] The pre-immigrant past was transformed and the gaucho elevated as the potent and unifying emblem of the new Argentine man. Some cosmopolitan intellectuals objected to the new romanticism, reminding others of the Argentine horsemen's "barbarism," untamed independence, disdain for authority, and allegiances with Indians and strongman caudillos. But as Argentine society became more and more foreign and heterogeneous, the mythologized gaucho appealed to large portions of the population and even to intellectuals and government officials who themselves identified with metropolitan values.

In a country lacking national heroes, the gaucho was the best candidate for a mythical one, a standard (if unrealistic) against which children of immigrants could orient their own values in opposition to those of the Old World of their parents. Nativist reformers held up the gaucho as an ideal national type, of *argentinidad*, as masculine, of the land, and close to the nation's "native" and traditional roots. The famous Argentine writer Leopoldo Lugones heralded the gaucho as the "hero and the civilizer of the pampas." Prominent authors and essayists such as Manuel Gálvez and Rojas incorporated these ideals in the national education program. The best-known celebrations of the mythical gaucho were José Hernández's *Return of Martín Fierro* and Eduardo Gutiérrez's *Juan Moreira*, both published in 1879, which captured in poetry and prose the gaucho as the standard bearer of *criollo* (native) Argentine identity; both became required reading for schoolchildren.[64]

Cultural nationalism was, on the surface, a reversal of the dominant elite's ideology that had looked abroad for new citizens and for models. After 1900, the growing nativist movement urged traditional values, by which it meant Hispanic, Catholic, and conservative, eschewing the Nordic models that had been slavishly followed in recent decades (while holding on to the pursuit of European-style progress, prosperity, and modernity). Cultural nationalists such as Manuel Gálvez lamented the "foreign" nature of the nation, with its "hordes of Italian peasants," Jews, and anarchists. He criticized imitation of northern European culture. Reflecting the shift to what elites saw as patriotic habits, the legislator Cayetano Carbonell wrote lyrically, "What is even more fundamental, grandiose, and human: the frank and spontaneous sentiment for the country of adoption, the one of his children, that he is attached to, his fortune; or the only flag that has guided him on his life path?"[65] In a critique of Juan Bautista Alberdi's famous founding credo "to govern is to populate," politicians and intellectuals maintained that the goal was now to "Argentinize."[66] Nationalism was by 1910 the dominant sentiment among the political elite.

TEN

"Fully Attacking the Source of Moral Infection"

Purging the Nation of Incurables

> The Chief Executive will order the departure of all foreigners whose conduct
> compromises the national security or disturbs the public order. . . . The
> foreigner against whom the order of deportation has been decreed will have
> three days to leave the country.
> —"RESIDENCE LAW," Law 4144, Articles 2 and 4

On May Day 1890 the first public collective action by workers occurred in
Buenos Aires. Against a backdrop of economic crisis, first railway workers,
then the city's bakers, and finally factory workers and day laborers abandoned their places of work for the streets, where they clashed with police.
The 1890 event inaugurated the decades of general strikes and mass demonstrations that would become a dominant feature of life in Argentina. Work
stoppages and street violence escalated sharply between 1902 and 1910,
organized by anarchists. Vastly popular among workers at the time, the
anarchists' strongest group was the Argentine Regional Workers' Federa-
tion. At its peak, the FORA (Federación Obrera Regional Argentina) had
20,000 members, about 5 percent of the city's population, but it represented more power than its numbers, especially during times of economic
crisis. Calling for general strikes, demonstrations against police actions, and
protests against repressive and xenophobic state actions, the FORA inserted
into Argentine political culture a new element of militancy.[1] Unions or
political groups would post announcements in left-wing papers such as *La
Vanguardia* (The vanguard) or *Voz del Pueblo* (Voice of the people) that
workers would be gathering at a certain location, perhaps in front of a
factory, warehouse, plaza, or mutual aid society, on behalf of a specific
cause: protesting an employer's unfair treatment of workers or the firing of
a compatriot, demanding better working conditions or, most often, higher

wages. Crowds of up to 5,000 would gather, filling the streets and building support as they went, sometimes ending in violent dispersal by police. The capital's two major newspapers, *La Prensa* and *La Nación*, reported work stoppages and riots and the ensuing conflicts between police and demonstrators.[2]

As the much touted May 1910 centennial of independence from Spanish colonial rule approached, aggrieved labor saw an opportunity to broadcast their issues widely in the national press, even beyond Argentina's borders. Protests were organized, and rumors of planned violence and sabotage reverberated throughout the city. Tensions mounted as police and government officials, anticipating a class war in the streets, launched a preemptory crackdown to prevent any embarrassing disruption of the celebrations. By many popular and leftist accounts, the state's preventative actions were unduly harsh. Municipal and national authorities, marshalling the forces of defense, had one important tool at hand, growing police power, and they were willing to implement it. They bolstered military resources as well and set up special public exercises to display the armed forces' presence and authority. The president declared a state of siege, an incongruent move for the birthday of the nation's liberty. The May centennial celebrations, fortified with police and curfews, passed with a minimum of social disruption. A little more than a month after the peaceful events, however, Buenos Aires was shocked by an exploding bomb at one of its cultural gems, the elegant Teatro Colón, just blocks from the president's palace. On this night of 26 June fourteen people were injured. Police pointed the finger at a Russian immigrant anarchist named Romanoff, although anarchists suspected authorities of looking for an excuse to revive the state of siege and to accelerate immigrant deportation and other repressive measures.

In reaction to these events, the nationalist movement spawned a new form of extremist right-wing vigilantes characterized by a virulent—and growing—strain of cultural nativism, xenophobia, and anti-Semitism. These civilian groups of upper-class city residents, mostly young men, most but not all of *criollo* (native) descent, were angered by the threat of working-class disruption of a glorious May 1910 celebration of the nation's promise and progress and set out to punish the rebels. Serving to bolster official state efforts to maintain order, about 1,000 of these vigilantes, together with conservative Catholic and women's groups, marched through the streets singing patriotic songs and parading the national flag, heckling bystanders who did not show the proper respect, and tearing down any Brazilian flags in sight. (Brazil had failed to send delegates to the official centennial celebration.)

After the bombing of the Teatro Colón, new paramilitary groups, who represented joint work between the police and the army, joined the youth gangs and, under the guidance of police chief Luis Dellepiane, plundered shops and offices of workers' newspapers, beat and sometimes killed "Russians," whom they associated with anarchism, and raped women.[3] Their special target was the Jewish neighborhood *Once*, where they attacked its cultural center and library and burned the books. The very next day, Congress approved its most xenophobic and repressive legislation to date: the Social Defense Law, an amplification of the Residence Law of 1902 that provided police additional powers to suppress revolts. The measure enhanced police power significantly. Police could now round up and deport dangerous foreigners. They could ban public meetings by anarchists and other so-called enemies of the state; newspapers and propaganda materials could be confiscated. Punishments for terrorism were strengthened. Most important, the law assumed that the most urgent dangers confronting the nation were of foreign origin and that the nation was polarized and divided. In the debate over the Social Defense Law on the floor of the legislature, Deputy Manuel Carlés warned, "We should not be surprised by the bomb explosion [in the Teatro Colón], since it appears that modern progress brings with it, as a consequence of civilization, that work of barbarism."[4]

Elected leaders and state bureaucrats, set to reclaim the nation from foreign barbarous forces, applied authoritative scientific language and concepts to political repression and social control. A legal program of hygienic purging of the nation was underway, whose focus was not only to prevent terrorism and political crimes but also to manage the nation's "germ plasm"—an approach that had both sexual and racial implications. In the words of one deputy, the best outcome of legislative debates about controlling anarchism was "laws to create healthy and strong nations."[5]

"Preventive-Social Hygiene": A Permanent Segregation of the Dangerous Element

The guiding spirit of the legal measures to ensure public order was "social defense," a central concept of the Italian school of criminal anthropology. Coined by the Italian criminologist and politician Enrico Ferri, the term described a system of "preventive-social hygiene" and was an integral component of that school of positivist criminology. Seemingly a natural outgrowth of the scientific study of crime, it asserted that the protection of society is the ultimate goal of criminal justice. Ferri's program stressed the importance of law and penal reform but also made suggestions for social reform, including rehabilitative punishment and special treatment for the criminally insane.[6]

In Argentina, "social defense" was interpreted by one criminologist in 1913 as justifying quarantine from society—the "initiative to segregate permanently from society those beings whose presence constitutes a menace to public safety, taking into consideration secondarily the seriousness of the crimes committed." Through the application of new scientific approaches, the criminologist continued, such criminological ideas had "already in large part penetrated" legal institutions.[7] Similarly, O. Gonzáles Roura, a provincial judge, wrote in the *Archivos de Psiquiatría, Criminología, y Ciencias Afines* (Archives of psychiatry, criminology, and related sciences), "In principle, crimes are public; they are only exceptionally private. By a general rule, then, they affect the collective interest, the public order, and social security." Roura expanded the definition of social security to include "peaceful homes, family security, social integrity, and the majesty of law."[8] Scientists draped their ideas in the powerful rhetoric of national security and insisted that their science alone could provide the best formula for protection of Argentina's precarious economic and cultural gains.

In 1909 Argentina's leading criminologist, José Ingenieros, wrote that the basic function of the law was to protect society from the base instincts of the individual. Crime, in that sense, was above all a violation of the "limitations imposed by the collectivity on the individual in the struggle for existence."[9] According to Ingenieros, "social defense is the rational base of a scientific system of punishment, expressly calibrated to the dangerousness of the offender."[10] Moral codes had been replaced with a formula for collective success in the chaotic Social Darwinist reality of society. Spanish criminologist Pedro Dorado commented that "without the law . . . peaceful coexistence and social cooperation would be impossible, and for this reason [laws] are usually also called the means of defense against the internal enemies of each society."[11] Similarly, author Francisco del Greco defined the "typical criminal" as "a man who lives in dissidence with the social group in which he agitates." Del Greco expanded the definition of criminality to include any break with social conformity: "This dissidence implies a rupture in the links of tradition, beliefs, and customs that include the individual in the social group. From that immoral conduct is born and it assumes a criminal character when it violates the group's fundamental conditions of life."[12] The social pathologists believed they had special instruments to offer their society in solving the spiking crime rate. Ingenieros wrote that "the correlation between penal criteria and our classification of criminals facilitates the practical applications of criminology, reconciling clinical and legal measures in order to transform criminal justice into an institution of prophylaxis and social defense."[13]

A 1913 draft criminal code recognized the tension between the need for order and the right to liberty, a dilemma especially acute for Argentina, a "young and extensive" country that was "inhabited by a cosmopolitan population, migratory and nomadic . . . a country in which the urban centers as seats of legal authority are spread out over enormous distances and where the police, even at the lowest levels, are required to obtain permission in order to contain criminality."[14] New crimes against the public order went beyond standard matters of criminal activity to include public threats, membership in illicit organizations, and crimes against public health, public order, and national security. They supplemented the earlier criminal codes' offenses, like the poisoning of public water systems, adulteration of pharmaceuticals, and deliberate spread of contagious disease, such as venereal disease. The 1904 draft, for example, introduced a few new categories, among them disclosing political or military secrets and causing harm to the national interest in international negotiations. Other offenses against "good customs" [buenas costumbres], like public drunkenness, corruption of minors, disrespect of women in public, cruelty to animals, and gambling, were punishable by fines. It all added up to a mounting preoccupation with social order.[15]

Of the "dangerous" men in society, anarchists were deemed the most alarming to the elites of Argentina. In the words of one police report, anarchists were the "source of social pathology, inassimilable to our collective character" and displaying an "exotic atavism."[16] They carried germs of different kind—social and political ones that could infect the minds of impressionable workers. Anarchists had been vilified ever since the first ones had arrived in Argentina in the 1890s. An 1897 article by Francisco de Veyga, the hygienist turned criminologist, termed anarchists an "imminent and continuous danger to the social order." Veyga warned that "no society can resign itself to tolerate them coming into its midst to unceasingly conspire and realize their plans of destruction," and he called for increased police and legislative efforts based on scientific strategies like the isolation of vagabonds and "parasites," reclusion of criminals, and sequestering of the mentally ill. He wrote of the anarchist: "Fully attacking the source of moral infection where these virulent germs sprout, we would destroy the dangerous element that is destined to produce political crimes." Veyga pointed to the encroaching violence and "repetition of anarchist attacks" as a reason for scientists to "occupy ourselves with these deeds and the factors that produce them."[17]

As early as 1882, the Buenos Aires police had begun to compile nationality-specific records for arrests and incarcerations, which reflected the belief that immigrants were responsible for disorder.[18] Calls for the deportation of immigrant anarchists became louder and more frequent. The first such legal proposal, written in 1899 by lawyer and diplomat Miguel Cané, was bluntly entitled *Expulsión de extranjeros* (The deportation of foreigners). It demanded the removal of convicted criminals and "exclusion of all foreigners whose conduct could compromise national security, disturb public order or social peace." Those individuals to be deported were to leave within three days of their sentencing.[19] *Expulsión del extranjeros* was essentially Cané's expanded, "scientific" version of existing legal concepts on immigration standards. The daily newspaper *La Prensa* published Cané's proposal in its entirety, as well as commentaries on it.[20]

While the concept of deporting "undesirables" had long existed in Argentina, Cané, in 1899, had introduced a further element of urgency that fit well with the spreading nationalism and antiforeigner sentiment in the legislature. He considered deportation "inherent to national sovereignty" and "legitimate and universally recognized," and the proposed law "fully constitutional": "All the European nations have it in their legislation and constantly exercise it." Cané cited as Argentina's "constitutional model" U.S. immigration laws, especially the restrictive measures of the period that were aimed at Chinese immigrants. To Cané, immigrants were the principal source of Argentina's social problems: "Our country is the promised land for all European vagabonds and delinquents." "Few trials of anarchists," he said, "do not reveal the presence of some of the accused in Buenos Aires or some other Argentine city." To give the state the power of expulsion would guarantee a "healthy" immigration and "foster peace." Cané wrote, "I sense a certain discomfort, not only in the political body of the State, but also among the people—nationals and foreigners alike—as a result of the absolute lack of defense encountered in this country against the elements of perversion that constantly arrive."[21]

In 1902, with the nation rocked by violent general strikes that were frequently blamed on the growing anarchist groups, representatives at a special session of the legislature declared as part of the Residence Law measures enacting a state of siege to end the strikes. Interior minister Joaquín V. González proposed the law "because of the strike declared by numerous workers from different unions who threaten the public order, our commercial and shipping interests, and therefore our public wealth."[22] The anarchist events of the early 1900s were tapping into one of the Argentine elite's deepest anxieties: that their nation's progress and modernity would be slowed or even reversed by social turmoil.[23] Deputy Ramón Vivanco considered the Residence Law of 1902 justified in the context of "civilization" for Argentina, as for other "advanced" countries: "All the clauses of this project form part of the legislation of civilized nations." Vivanco's colleague Juan Angel Martínez invoked modernity, praising the president for "the very prudent and discrete use of these laws," which had been carried out "with commendable equanimity and worthy of the leader of a civilized state." While the 1902 proposal had encountered opposition, mostly among those who were concerned about its constitutionality or among those who represented immigrants' interests—socialists, for example—the 1910 Social Defense Law was passed with little opposition. Promoters of this piece of repressive legislation used the United States as a model of "progress" and of seemingly effective anti-immigration policy. The United States, González noted, had faced similar constitutionalist criticism but had resolved the issues. The constitution, he argued, was intended to defend the country "against disorder brought from abroad and that are outside of the constitutional mechanism."[24]

Even opponents of the 1902 Residence Law also regarded social conflicts as an inevitable outcome of Argentine progress. In that vein Deputy Rufino Varela Ortíz, skeptical about the proposed law, called strikes "a phenomenon of modern society," though he pointed out, without effect, that the leadership of the anarchist movement was in the hands of native-born Argentines, not of immigrants. Deputy Emilio Gouchon defended the workers by invoking Social Darwinist concepts. "Until now," he said, "strikes have not been a bad thing. Strikes have been the legitimate defense taken by the working man against the capitalist." Yet Gouchon, too, lent his weight to "social defense": "The laws should be adjusted to the social alarm, to the social evils that produce those acts."[25]

Criticism of the 1910 Social Defense Law revolved around constitutional issues, but even critics maintained that society had a responsibility to defend itself from harmful elements in its midst. In 1911, a year after implementation of the law, legal scholar Francisco Durá published a critique of the law that rested mainly on constitutional grounds: the law contradicted the guaranteed rights of individual residents of Argentina. The legislation had been passed in "two situations of public panic" and had not been sufficiently deliberated. Durá himself did not doubt the nation's need to defend itself against terror and chaos, but he wrote, "To defend society in any circumstance and against all bad elements, it is not necessary to contradict the Constitution with laws. . . . To defend the nation dictatorships are not necessary. . . . [In] the Constitution, we have an honest starting point for laws that screen foreigners on their entrance into the Republic, that regulate the effects of their civil residence, and that permit the country to expel as poisonous those that really are."[26]

Durá's suggestions called for a graduated scheme, with a two-year probationary period during which any immigrant who committed "ordinary crimes," any anarchists who proselytized, or anyone exhibiting signs of "moral corruption or physical defects or disability" could be expelled. After two years without a transgression, the immigrant could be deported only as part of sentencing for a crime, or in case of war or domestic conflict "if the foreigner does not obey Presidential orders to move." (In 1908 legislator Lucas Ayarragaray had suggested a ten-year probation period, easily double the amount of time proposed by most legal scholars; see chapter 9.) Durá also proposed as a legal complement the revision of the criminal code to require deportation for certain "defined, classified, and technically and specifically considered" crimes, among them repeat offenses, corruption of minors and sodomy, profiting from the "white slave trade," provoking or leading a strike, committing violence during a strike, and issuing anarchist propaganda. Most important, Durá recommended the deportation of "the *born criminal* . . . [as] he is presented on a daily basis with the complete inefficacy of the laws."[27]

Between 1902 and 1915, 383 immigrants were deported, the majority of whom were Italian or Spanish.[28] The impact of the new legislation on the nation's political culture had sent a powerful warning to the entire population in defining—with the backing of the law—the boundaries of acceptable and unacceptable behavior.[29] Other individuals were confined but not deported, with most arrested anarchists sent to the National Penitentiary. The worst offenders were sentenced to internal exile in the remote prison in the coldest part of the country, the "Argentine Siberia," at the tip of Tierra del Fuego, a dreaded destination reserved for the hardened, untreatable, and irredeemable—the "born criminals." The prisoners held there were sentenced to hard physical labor and forced to build what became the small town of Ushuaia.

The rhetoric in the legislative debates on the social defense measure had been vitriolic, since the political elite regarded the growing anarchist movement as the primary threat to Argentine civilization at the turn of the century. Legislators saw anarchists as a separate and inferior category of human being. Conservative deputy Lucas Ayarragaray, for example, wrote that anarchism "is constituted by a band of degenerates and fanatics." Physicians linked anarchism with the atavism and primitivism of the crowd and with mental and biological degeneration. Another deputy called for repression of all anarchists regardless of nationality. "It is not necessary to declare anarchists outside of the law; it is necessary, on the contrary, to locate them under the power of the law. . . . [T]he author of the most repugnant of all crimes [is] the traitor of his country!" So hated were anarchists that most on the right and the left viewed them in similar terms. Although Socialist deputy Carlos Meyer Pellegrini defended their rights, he simultaneously referred to anarchists as "monsters" and "human beasts." Legislators shunned anarchists as "cowards," as men who rejected legitimate work, and questioned not only their patriotism but their masculinity. Deputy Manuel Carlés, a noted author on social problems, said that the Teatro Colón bombers showed in their behavior "the perversity of the intent and the cowardice of the action." Employing these concepts, legislators were communicating ideals of citizenship, of who could and should participate in the public sphere. Civil responsibility and political participation, good and bad, were limited to men willing to follow ideals of behavior. Ayarragaray agreed: "We believe that the family and property are the definitive and irreplaceable bases of all civilized organization."[30]

The new punitive legislation was intended to eliminate from society miscreants like Simón Radowisky, a young Russian immigrant and anarchist who shared the popular sentiments of hatred toward the government, especially the city's police. On a clear day in November 1909, Colonel Ramón Falcón, chief of the Buenos Aires city police, was paying his respects at the funeral of his distinguished colleague Antonio Ballvé, former director of the National Penitentiary, when, as Falcón's carriage pulled away from the cemetery, Radowisky hurled a bomb through the window. The chief and his secretary, Alberto Lartigau, were killed. Radowisky was instantly seized and jailed. Anarchists themselves were convinced that Falcón had personally instigated recent events of police brutality against demonstrating workers. The assassination of Falcón did not bring an end to police violence but merely fanned the smoldering class tensions that had plagued the city's everyday life for the past two decades.

The capture of Radowisky made him available to the city's medicolegal and criminology experts as a physical and psychological criminal subject.

Physicians were determined to discover the psychological cause of Rado-wisky's aberrant act, and hence to prevent other such acts, and also to settle on his age and degree of responsibility for his crime. The court verdict in the trial of Radowisky was published in the nation's main criminology journal, *Archivos*; according to this account, Judge Sotero F. Vázquez had applied the latest criminological categories to his assessment of the deed. He concluded that Radowisky was "not a degenerate or a neurotic; he is a normal subject who has committed a crime in full control of his faculties."[31] With respon-sibility for the act established, professionals of the criminal justice system could now begin, "scientifically," to determinate appropriate disciplinary measures against Radowisky and to search for ways to avoid similar crimes. After serving the first two years of a life sentence in the National Peniten-tiary (he was spared the death penalty since he was finally determined to be a minor at the time of the crime), Radowisky was transferred to internal exile in the Ushuaia penal colony. But, over the years, Radowisky became to many a national hero. His fate was seen as a sacrifice, epitomizing the harsh and punishing approach to law and justice in Argentina in these years.

With the Falcón-Radowisky tragedy firmly in mind, legislators set out to enact new laws. Opening the legislative session on 27 June 1910, Deputy Eduardo Oliver introduced legislation to "prevent hordes of criminals from arming the abject and the lost [and] from arriving and through terror de-stroying the social order and social organization." The "monsters" who would murder "elderly and defenseless women and innocent children" were "outside of the protection of all social laws." Even Socialist deputy Meyer Pellegrini called "this evil of anarchism" a "sickness that dates many years back . . . that, imported by a few could be isolated and cured in principle. But today, unfortunately, this evil is so dispersed in our country, that it would be impossible to demand that the legislature decree a law that would cure it in one swoop."[32]

Medical concepts such as contagion, degeneration, and biological stigma permeated the legislators' discussions of anarchism. Lucas Ayarragaray, himself author of a widely reviewed book on Argentine socialism and the working class, now characterized anarchist terror broadly as a "monstrosity that today germinates in the hard heads of some proletarians provoked by the sick declamations of certain ideologues." On the floor of the Senate in 1910, he warned that "new needs and new forms of crime have emerged, situations without precedent, which have created revolutionary anarchism and socialism. We thus must break the recognized molds, create adequate legislative forms in concordance with the new ideas and deeds that have provoked social and economic complications . . . and have given birth to ferocious aberrations in the European proletariat." Ayarragaray's colleague Adolpho Mugica concurred; acts of terrorism were "simply the product of criminal instincts which were hatched and born in an environment distinct from ours . . . precisely because here they find the tolerance which does not exist in their countries of origin." The legislators called for "scientific studies" to prevent future attacks.[33]

Ayarragaray also proposed a modification of the immigration law to bar from naturalization immigrants who had been previously sentenced for crimes in Argentina or in other countries, as well as all individuals deported from any nation for anarchist activity.[34] Anarchists provoked the "scorn of civilization and the scorn of our laws," Ayarragaray argued; "I will not let their ignoble red rags pass by our blue and white flag that symbolizes the glories and efforts of a great nation. . . . Anarchism has developed among us . . . from a lack of authority and contrary counterweights. Let us prevent the anarchist delirium from spreading to society's lower classes and taking on barbaric form." But the attack went beyond anarchists, with Ayarragaray's adding the mentally ill, epileptics, and the insane as targets of the new law. The anarchist, through the use of the press and other forms of proselytization, "works to impress and gather followers, but among the mentally weak and predisposed is where the idea of the criminal deed makes inroads, where one can find the degenerate who throws the bomb."[35]

Both the Residence Law and the Social Defense Law articulated a fear of threats not only to the social order but to *argentinidad* itself. The Residence Law was necessary, Ayarragaray said, because it "permits the exclusion of those foreigners who come to disturb the social order and ruin Argentine laws, with foreign elements that disintegrate our character and our history." He deployed concepts of nationalism and national identity to purge poisonous outsiders. His views were widely reflected among colleagues in the legislature. One deputy demanded the passage of the Social Defense Law "to defend the Republic against those advances of bestiality and crime!" Another described the fight as a war of the Argentine "people" against a foreign enemy. "We cannot say to society that we cannot give it efficient means of defense, nor permit it to defend itself! Why? Because this is to deliver society into the hands of anarchism!"[36] Invoking the language of crowd psychology, Deputy Juan Balestra wrote that "the attacks of the striker against those who do not accompany him, these acts of pressure and contagion, are those which constitute the phenomenon of the strike."[37]

Upon its passage the Social Defense Law, building on the Residence Law,

provided that "the entrance and admission of the following classes of foreigners will remain prohibited . . . anarchists and other persons who profess
or plan attacks by any means of force or violence against public functionaries or officials in general or against social institutions; [and] those who
have been deported from the Republic. . . . Should the culprits of the crimes
previously cited be natural born or naturalized citizens, in addition to their
sentence, they will lose their political rights and retire their Argentine citizenship." The new law also expanded police powers, limited the right to
assembly, and authorized the closing of newspapers deemed subversive.[38]

Left-wing newspapers such as the socialist weekly (after 1905, daily) *La
Vanguardia* spoke out against the states of siege and antiforeigner legislation, characterizing police actions as "gags" on the people and "against the
workers' movement and the progress of the nation." A 1905 authorization
of a state of siege, one headline proclaimed, had been "defended in the
chamber by a drunken deputy."[39] The Residence and Social Defense Laws
were even more of a travesty, the papers argued. This "barbaric" Residence
Law, *La Vanguardia* said, amounted to nothing more than "low police vengeance . . . dictated in a black moment by a congress of eunuchs, for the
calm digestion of the bourgeoisie."[40] The anarchist paper *Voz del Pueblo* took
a skeptical view toward the definition of foreign workers as "dangerous":
"There are much more dangerous individuals than those who form the
nationalist clubs [i.e., foreigners]. The people in the brothels who elect the
authorities who rule us—from the legislature on down, including the governor and the vice governor—and the clean Catholics who warm their pews;
[they are] more dangerous than the migrating political activists." The only
crime of the foreigners who demanded better living and working conditions
was that they "have ideas, they are a little proud." The article remarked
sarcastically, "Ah! That's the sin! They are men! That is not permitted to
foreigners . . . nor to the sons of the nation either." The article concluded by
taking on the voice of imaginary conservative legislators: "Get out, then,
you dangerous foreigners, you are disturbing our *siesta*!"[41] The short-lived
labor newspaper *Vida Nueva* (New life) also took to mocking the establishment. Under a front-page headline, "Total Tyranny," it placed a large por-
trait of a monkey in fine clothing with a newspaper on its lap. "The legislator who voted for the Residence Law," read the caption.[42]

Despite such criticism, proposals to deport dangerous foreigners made
their way into draft codes as well. A 1913 version stated, "In an immigrant
country" such as Argentina, among the "thousands of foreigners who arrive
on her shores to work . . . wrongdoers slip themselves in and take refuge."[43]

Three years after the successful passage of the Social Defense Law, penologists at the Argentine Penitentiary Congress heard extensive commentary about the law by the judge O. González Roura. He flatly stated that "all of Law 7029 . . . will remain included in the draft criminal code [of 1913]."[44] One uniform code for the country would simplify the policing of crimes.

Legislators who considered Argentina a country generous to immigrants and full of opportunity for advancement saw anarchists as defying a social pact laid out in the Argentine Constitution. Ayarragaray justified this view with a distinction between "the foreigners who come to populate and work the Republic's earth . . . and those who arrive on our shores, if not with the mark of the European prison on their body, with the sentence that condemned them to an errant life, away from their homeland."[45] Despite avant-garde scientific methods, the ability to distinguish between desirable and undesirable members of the nation rested on age-old expectations about social worth and productive labor.

Perfecting the Social and Legal Order: Fingerprinting as Prophylaxis

Despite public criticisms, the social defense laws were immensely admired by officials and merely codified views that had been gaining in legitimacy for thirty years. Police strongly supported the laws (see chapter 6). As early as the early 1880s a capital police report had firmly linked foreigners with crime. The report agreed with current theories that the solution to this problem was assimilation and the "education of the masses."[46] Social control, by force if necessary, was clearly essential to this effort. The capital police positioned themselves as crucial to the nation's advancement by promising to maintain public order and discipline the unruly crowds of immigrants. Supporting these sentiments, the *Revista de Policía* (Police review) reprinted in its entirety Miguel Cané's proposed 1899 law on the expulsion of foreigners.[47]

The capital police had focused their attention on the immigrant crowd as harboring criminals and troublemakers. Calling the 1902 Residence Law a "law of public health," the police now suggested its expansion to control vagrancy and other quasi-criminal social disturbances "in order to provide better and more healthful effects."[48] Fingerprinting was seen as a complement to these measures, a promising way to capture criminal immigrants. In 1911 Francisco Durá, an influential legislator, proposed modifications to the existing laws on immigration, exclusion of foreigners, citizenship, and naturalization that would require "the most perfect personal identification of the immigrant," specifically "the taking of fingerprints." To aid social defense efforts, a copy of a new arrival's fingerprints would be dispatched to the immigration office and a second copy to the chief of police at the immigrant's destination.[49]

Buenos Aires city police chief Luis J. Dellepiane provided statistics in his 1912 annual report that demonstrated a rise in the number of disruptive strikes and incidents of street violence. He looked to a repressive law for a solution, citing, in particular, the pledge of the Social Defense Law to mitigate the "inevitable conflicts between capital and labor."[50] The new chief, Eloy Udabe, wrote to the interior minister that the legislative measures were not enough: "Vagrancy, begging, pederasty, [and] abandoned children are social plagues that my predecessors have made known in many communications and proposals sent to the highest government," he said, in urging new measures to provide for the "prophylaxis of social sanitation." Udabe, however, singled out anarchists as the most serious and immediate peril, noting that 386 individuals (109 of them Argentines, 277 foreigners) had been arrested for infractions against the Social Defense Law. Anarchism in its semiclandestine role demanded serious vigilance. Of this hidden danger he wrote, "its criminal tendencies manifested in the components of the [anarchist] sect, always latent, can at any moment surprise society, such as with the concrete example of the personal attack."[51]

Crucial to the implementation of the Residence and Social Defense Laws was a thorough reorganization of the police. Under the leadership of police chief Ramón Falcón in 1906, the force had become more militarized, with special sections devoted to social unrest. The capital police's investigative division took charge of social policing, including the maintenance of public order, "by means of knowledge, preventative observation, and repression of socially disruptive elements" and for "classifying persons reputed to be dangerous to the public order or with criminal habits."[52] The police needed, so they believed, a "magic bullet" to fight anarchism and socially disruptive crime. Fingerprinting provided such a technique for dealing with the effects of the dual pressures of rapid modernization and of mass immigration. Though the procedure originated in Europe, it first emerged as a tool of the state in the developing bureaucratic system of Argentina, because it spoke to the specific needs of that heterogeneous immigrant, "anonymous," and peripheral society.[53]

The success of fingerprint identification in ridding Argentine society of its dangerous elements was recounted in a 1913 article in the *Archivos*, which described how the capital police solved a robbery case with the use of fingerprint techniques; the investigating detective included in his report the recommendation that the perpetrator, who operated under a number of aliases, be deported under the letter of the law.[54] Fingerprinting's advocates began to apply the process to nearly every realm of human experience— marriages, births, and deaths; public testimonies, legislation, and contracts; public health, welfare, residency, prostitution, and vagrancy. Luis Reyna Almandos, a colleague of the Buenos Aires police scientist and fingerprinting innovator Juan Vucetich, evoked the concept of social defense in explaining the technique's value, writing in 1909 that "Vucetich's system tends to consolidate the social order in all its aspects."[55]

Vucetich, who devoted his professional life to promoting the use of fingerprints for both criminal and civil uses, attempted to influence the new proposed criminal codes of 1906 and 1917 by stressing the need for an updated procedure for dealing with repeat offenders. He described the 1906 draft code as "a legal measure which for the first time in world history incorporates in its provision numerous articles based on personal identification by means of fingerprints."[56] The authors of the draft code had included thirteen articles involving fingerprint technology, which Vucetich noted were not to be found in any previous law. He pointed out approvingly that the 1917 proposed code also dealt with the "grave problem" of recidivism, and he cited the authors' linked concerns about recidivism with national interests:

> Up to the present, recidivism has been considered as if the Capital, the national territories and the provinces were independent states with regard to criminal law. . . . So far as the Penal Code is concerned, the Nation is one, and consequently all who have committed a crime before must be considered previous offenders, whatever the national territory was or whatever court imposed the sentence. . . . Therefore, if there are no provinces or territories and if there is no capital, but merely one Nation, under the Penal Code, this reform is justified.[57]

In 1908 Nicéforo Castellano, chief of the Office of Identification at the capital police, published a draft law to create a National Identity Office.[58] Citing the international recognition of the Argentine fingerprint system, Castellano stated that the system had been "accepted in spite of the handicap of its being an American invention. . . . This young country should begin to exploit the advantages of this progress by establishing practical applications from it."[59]

To this end, one scientist, writing in 1909, proposed that fingerprinting become the centerpiece of a "General Register of Prostitutes," its purpose to reduce the incidence of syphilis.[60] Reyna Almandos, too, suggested that prostitution, though an "incurable evil, since it is inevitable," could be controlled by fingerprint science. In the fight against syphilis, a dactyloscopic system of identifying prostitutes would indeed be the best weapon: "An identified woman will remain definitively impeded from using different names, so that she cannot deceive the vigilance of the sanitary authorities." Reyna Almandos further recommended a registry for domestic servants to prevent the supposedly common occurrence of stealing and deception of their employers. He made a similar argument for the sanitary efforts against the "white slave trade," vagrancy, and begging.[61]

The police, while most interested in immediate uses of the method in apprehending repeat offenders, were aware of its wider implications for social control beyond the criminal world. Dactyloscopy showed promise, as one police official mentioned in 1909, in its "multiple and diverse applications."[62] The ideas circulating in criminal justice circles were heard in government administration as well. That same year, a provincial legislative deputy, Octavio R. Amadeo, submitted a proposal to create an identification register based on Vucetich's system.[63] Argentina became one of the first nations to take the fingerprints of certain sectors of its civilian population. Police force candidates were fingerprinted beginning in 1891, and individuals applying for other government positions in the early 1900s, postal employees in 1910, and students and other governmental staff in 1911 were all fingerprinted.[64]

In 1914, in an expansion of the police reach, Vucetich, Reyna Almandos, and other scientists attempted to broaden the application of fingerprint technology to the civilian population. They proposed establishment of an "Argentine Dactyloscopic Institute" on the model of the Pasteur Institute in France. The founding statement of the institute claimed, "If the goals of the Pasteur Institute tend toward the improvement of physical health, those of the Dactyloscopic Institute tend to perfect the social and legal order of cultured nations."[65] A committee, chaired by Reyna Almandos and consisting of six other judges, academicians, and government officials, was formed to raise both private and public funds for this "patriotic endeavor [that is] called to lend incalculable services to our country and to the civilized world." Jorge A. Susini, the commission's secretary, sent hundreds of letters to newspaper editors, bank executives, and corporate chiefs of large companies. He pitched his letters to each specific audience; for example, he pointed out to business leaders the value of fingerprinting to the discovery of fraudulent financial operations.[66] Reyna Almandos added that their financial participation was central to the success of the institute, which he pointed out, would "fundamentally and beneficially influence the commercial, political, and social development of our lands."[67]

In seeking to establish the Argentine Dactyloscopic Institute, solicitations were also made to national and provincial officials by appealing to the government figures' sense of national progress. The commission hoped the delegates would "wholly contribute to the realization of the idea pursued by this Commission by mandate of the Assembly celebrated in La Plata on 28 November 1913." "We attempt, Mr. Delegate," the commission stated, "to establish with the help of the public and the Public Powers the formation of high scientific culture that will lead to the highest perfection of the Republic's civil, political, judicial, military, etc., institutions." Citing the success and "universal application" of Vucetich's method, the commission promised that the new institute would "produce a radical and beneficial transformation of public institutions in our country and abroad."[68] Similarly, a pamphlet published in 1914 by the fledgling institute cited Vucetich's role in demonstrating his method in Europe, America, and Asia not only "with the object of making the system known and implementing it, [but] at the same time giving them the exact idea of the progress and civilization of our country."[69]

The main goal of the Dactyloscopic Institute was to promote and lobby for new laws to require application of fingerprint science in civil, political, and administrative matters, as well as in criminal justice procedures. The organizers of the institute even proclaimed that dactyloscopy would foster legal equality: "Prejudice is dead, because the system is *egalitarian*: all are equal before the law and all are different before Nature. . . . There are not two leaves alike in the forest; there are not two men alike in the world."[70] They were convinced that the introduction of a personal identity card complete with fingerprints would "guarantee the right of identity that each of us has from the moment of birth." Those who would institutionalize fingerprinting among the general public evoked images of the crowd. "Each person carrying an identity card would immediately cease to be unknown, even among the immense and foreign crowd," advocates promised. Names and even faces can change, they maintained, but not fingerprints.[71]

The largest and most concerted effort to apply fingerprint science to civil society at large was the campaign for a General Identification Register and national identity card. Registers of various types had existed since colonial days, for example, for births, deaths, and marriages. The intent of a General Identification Register was to access the fingerprints of every man, woman, and child in the nation. Vucetich could barely contain his enthusiasm for this new universal civil procedure, and he was the first to volunteer his own fingerprints. In 1915 Vucetich came one step closer to realizing his dream of universal fingerprint identification when a draft law to create his General Identification Register was proposed. Article 4.10e stated, "Social security, morality, and public well-being have an interest in the vigilance of the individual fingerprinting of persons."[72] To support the measure, Vucetich referred even more explicitly to concepts of social order and control. He wrote that fingerprint records "will constitute an element of social defense and public security, impeding those who elude or deceive the law through the obscuring of identity. It will be a powerful means of discovering all possible ulterior crimes."[73]

In 1916 Vucetich published his *Commentary on the Draft Law Creating the General Register of Identification*, which outlined his expanded dreams of universal fingerprint identification for Argentine citizens and residents. Vucetich's innovation in this area was twofold: he advocated creating a centralized system of registration, one for criminals and one for the general population; and he also made fingerprints the central measure by which all records would be classified. His proposal included the expansion of identification to the civil sector of the province; creation of a general register and archive for records of both citizens and criminals; organization of the criminal statistics of the province; education of judges about the identification process; establishment of a school of identification; and finally, the institution of a museum, library, and journal to "organize the diffusion of the system."[74]

Vucetich couched his proposal in terms of the inevitability of national progress and scientific advancement. "This project," he wrote, "responds to a deeply felt necessity in public institutions, and is the natural consequence of the development and growth of dactyloscopic identification." His proposed law, building on fifteen years of previous attempts, was instead to unite the three strands of thought on general population identification and control. On the one hand, it would identify, classify, and archive information about each citizen in the nation and require a national identification card for each. On the other hand, it would create a special register for criminals and the insane (in conjunction with improved criminal statistics) in order to identify recidivists and other dangerous elements in the population. Finally, it would establish a special "public security" register to police individuals who threatened public morality, order, and "well-being."[75]

Vucetich, however, in applying his theories, had no intention of limiting their application to the provincial government. He sketched and printed forms for every type of civil activity, including the national identity card and the registration of students and employees. Among Vucetich's papers on the national register are a number of draft application forms for identity cards in which he outlined the information to be gathered from each resident. In these drafts he plotted separate forms for Argentines, foreigners, and naturalized citizens. Exemplifying the civil use of criminalistic techniques, all three forms included a section for a physical description, "distinguishing marks," and fingerprints. Like Vucetich, the proposed legislation for a national identification register pointed out its potential practical effect on protecting the social order: "It is not enough in doing criminal statistics to take note of the number of defendants. It is better to order and coordinate them, linking them with the identity of the accused. According to the system adopted in this draft, not only the name and description of the defendants will be known, but also the distinct phases of judgment, prison movement, and the more or less perfection of the judicial mechanism."[76]

If it was to be effective over the long run, Vucetich believed, fingerprinting would have to be universally and uniformly used; therefore, to spread his method throughout the "civilized world," he corresponded regularly with police officials, scientists, and government officials in Europe, the Americas, and to a certain extent, Asia. He made a world tour in 1913, attending many international police conferences, where he shared his scientific method and his zeal for fingerprinting as a means of social control for criminal, civil, political, and administrative branches of government. He urged the establishment of "Intercontinental Identification Offices" to ease the exchange of fingerprint records across borders and to expedite the apprehension of criminals who were attempting to flee justice.

Despite all of Vucetich's hard work and his significant achievements, the concept of universal fingerprinting encountered opposition. Individuals did not always subject themselves willingly to anthropometric and dactyloscopic examination, and there was popular resistance, as well as judicial opposition, on constitutional grounds.[77] A national Office of Identification with fingerprint records of both criminals and "normal" citizens was, however, eventually established in 1933, eight years after Vucetich's death, and was charged with creating identity files and distributing a national identification card for each citizen.

"Errors of Sex": Social Prophylaxis and Sexual Control

The first decade of the twentieth century saw the emergence of a new form of racial policing and racial control. A central assumption of the liberal state builders at the turn of the century was that the country and its future citizenry would be built through immigration of the correct human "stock," a fair-skinned, fair-haired, healthy "European" population, that would contribute to the work ethic, morality, and "higher" culture. In 1910 Lucas Ayarragaray, on the floor of the national legislature, announced, "We do not need yellow immigration, but rather European fathers and mothers of the white race to improve [*superiorizar*] the hybrid and miscegenated elements that constitute the base of our nation's population."[78] This social danger required more than political purging. It had to be prevented at the source, in the constitution of the race itself.

Some Argentine elites worshipped northern cultures but hesitated to dismiss Argentina as a failure for its predominantly southern European composition. José Ingenieros, the country's leading psychopathologist, for example, asserted that there were no significant differences between white Europeans and southern Europeans, though he thought the latter faced more challenges since, so he believed, they had an element of African blood. Argentine historian Victor Arreguine's 1906 comparison of southern European and Anglo-Saxon cultures disputed the myth of northern European superiority over Latins (whom he defined as the French, Italians, and Swiss). Most great ideas, he said, sprang from the ancient civilizations of the Mediterranean, but their northern neighbors were adept at adapting and applying that knowledge.[79] A 1912 review article on "The Color Blond" cited a "disappearance" of lighter features among whites and an increase of dark-haired, dark-eyed people. The author of the review concluded that "the cause is confirmed: the constitution corresponding to light hair and eyes does not adapt to the conditions of city life, and cities are ceaselessly growing at the expense of the country."[80]

Racial ideas went beyond the selection and exclusion of immigrants to the control of the population as a whole on the biological and reproductive level. Social prophylaxis for the general population, including invasive tests, vaccination, and even sterilization and racial hygiene, all fundamental policies that would drive the worldwide eugenics movements at their peak in the 1920s through the 1940s, were gathering momentum in Argentina of the 1910s. Prominent physicians in Buenos Aires, including Victor Delfino and Enrique Feinmann, the founder of the field of puericulture, or maternal and infant hygiene, reflected the widespread concern with the quality of children born to immigrants. These physicians had previously encouraged racial selection and control of immigration at the Buenos Aires port of entry, sanitation in the slums, and a series of state interventionist efforts to minimize the harm done by mothers to their children through lack of education or innate intelligence. Fearing that fertility rates were dropping, physicians now also sought ways to increase the number of healthy births in order to provide future healthy citizens. With women seen as the unique carriers of the next generation, the subspecialty of puericulture took shape to impose "scientific" methods of mothering in an attempt to lower the appalling national rates of infant mortality and chronic disease.[81]

At the top of hygienists' list of concerns were venereal diseases, feared as risking the germ plasm of the nation by creating weak and deformed children and by raising mortality rates. Hygienists had regarded prostitutes the sole vectors of venereal diseases, and from 1876 on, prostitutes had therefore been the focus of coercive and obligatory prophylactic programs. Specifically, prostitutes had been compelled to register with the police, to live and work in specific areas of the city, and to submit to regular medical examinations (see chapter 5). Should a prostitute be found to be ill, she was hospitalized against her will. This system of control had persisted within a system of official condoning of the practice, even by the Catholic Church, until the 1930s, when prostitution again became a crime. The government now passed new legislation requiring prenuptial blood tests, this time for men only, since the earlier program of control had been unsuccessful. Sexual diseases were, in the words of hygienist Angel Giménez, "the social scourge of modern times." Syphilis killed 30,000 Argentines per year in the mid-1930s, and 40 percent of all hospitalized patients in Buenos Aires were there for venereal disease. Nearly 40 percent of the armed forces were similarly infected. Physicians believed that venereal disease in the parents could cause mental illness, blindness, polio, and paralysis in the children of prostitutes or in the children of men they had infected. The 1936 Law of Anti-Venereal Prophylaxis, as it was called, proposed by Giménez and another progressive physician colleague in the legislature, set up a National Institute for the Prevention of Venereal Disease to disseminate hygienic education and to regulate medications. Aside from the new mandatory medical certification for men registering to marry, the law required treatment of any person found to have venereal disease and criminalized the act of knowingly spreading the disease.[82]

Homosexuals and homosexual behavior were also the targets of hygienic efforts. Girls in boarding schools, women in convents, army recruits, and male prostitutes were all considered sources of social contagion. Physicians believed that homosexuality was yet another "decadent" by-product of modernity, but also that, because it was inadaptable to society, it would die out on its own. Psychiatrists who studied homosexuals suggested that the fragile psychological state of "inverts" would lead inevitably to high rates of suicide and to extinction of the type.[83] Hermaphrodites, too, were believed to be destabilizing to society. Physician Carlos F. Roche speculated that hermaphroditism's cause was "morbid heredity" or a lack of proper nutrition in utero. Either way, "one finds in the antecedents of these subjects degenerative stigma, either physical or psychological: a father, a mother, sometimes both, epileptic, hysterical, alcoholic, or with vices." The medical profession, Roche argued, should seek to eliminate or at least treat the abnormality by, for example, better nutrition for expectant mothers or prevention of degenerative births. Roche urged expert medical examinations, from "the moment of birth" through puberty, of persons suspected or confirmed as harboring abnormal sexuality. Most important, such persons should be examined "before marriage, with the goal of knowing if he is or not apt for it," and during marriage, if deemed necessary, "to seek to nullify it." Without strict surveillance and, when possible, medical or surgical intervention, "errors of sex such as these produce, and will always produce, grave disorders in society."[84] As prophylaxis for homosexuality Eusebio Gómez recommended, "Let us create a more integral and humane education in an attempt to destroy the atavistic strata from which these explosions of evil originate." Education alone was not enough, however, and he also demanded that the state's justice system be expanded to deal with this "evil."[85]

As in Europe and North America at the time, Argentine science offered proposals for sterilization and racial purging programs. One Buenos Aires writer in the journal *Archivos* suggested in 1902 that the state take up "artificial selection" of the population, "more efficient and quicker than natural selection, to be realized through the sterilization of degenerate individuals." Perhaps the most influential proponent of punitive eugenics was Joaquín V. González, lawyer and constitutional scholar, interior minister under President Julio A. Roca, and founding president of the University of La Plata.

González believed the state had a responsibility to prevent the reproduction of the unfit by implementing a science of selection to preserve the "race of tomorrow." He sought to replace the "inferior" races with "superior" ones through "that new science incorporated to the science of government . . . eugenic science."[86]

Most Argentine social scientists and public health reformers took a less extreme stance. They saw eugenics as a branch of hygiene, a way of ensuring public health, a progressive and innovative approach to social problems, and a challenge to the traditional stance of the Catholic Church by addressing sexuality and procreation in a "scientific" way. Like many Catholic critics, they found proposals to sterilize "degenerates" distasteful. (Proposals to sterilize the "unfit" were vociferously protested by the Catholic Church as "depraved and homicidal.")[87] Argentine scientists preferred limits or controls on immigration and racial typing over negative reproductive control. Some, emphasizing "preventative" eugenics, sought "progressive" solutions in public health programs through the control of alcoholism and venereal disease and in the improvement of living conditions for the poor.

The ideas of sanitation and hygienic engineering of public health had been discussed and implemented for decades, but not until 1918 did physician Victor Delfino first purposely apply the word "eugenics" in Argentina when he founded the Argentine Eugenics Society that year. A stronger eugenics coalition, the Social Prophylaxis League, would emerge three years later. Calls for state applications of eugenics in the 1910s and 1920s included proposals to limit immigration and to preserve the purity of the "race." In 1941 a prominent Argentine eugenicist, Arturo R. Rossi, described his work as dedicated to "defending white civilization" against the "profound polymorphism of our people."[88] Fears of racial miscegenation, in turn, fed the growing nativist movement in Argentina and provided a language and a confirmation of its belief in the necessity of preserving Argentina's racial purity along with its national identity.

A "New Civilization" Shaped by Science

Its Paradoxes and Its Consequences

> We find ourselves in the presence of a new social organization. Our life has
> moved enormously away from the past; it is not one of tyranny, of the forced
> expatriation of the most illustrious Argentines, and of a lack of citizens. . . .
> The political culture of the population has been raised prodigiously . . . [to] the
> level of any of the most civilized and ancient cities of the earth. . . . It is a new
> country, a new civilization.
> —CARLOS F. GÓMEZ

After the 1914 assassination of Archduke Ferdinand in Sarajevo and the
outbreak of World War I, Argentines were filled with a growing sense that
the French, British, and German values they had admired and emulated
were bankrupt. Criminologist José Ingenieros, watching the war in Europe,
called it the "suicide of the barbarians."[1] Argentina saw a new role for itself, to rise above its former models and to assume a position of superiority. The third national census of Argentina, published in June 1914, asserted that "the Argentine Republic should prepare itself to assume a role
that events . . . assign it, without doubt, after the European catastrophe. . . .
The great civilizations of antiquity disappear under the cataclysms of this
type, transferring from the East to the West; and it is not too bold to
suppose that . . . many factors of European civilization are being transferred
to our America."[2] Was Argentina now more "civilized" than its former
northern Atlantic models? Were *they* now the barbarians?

Optimistic, even boastful, assessments of Argentina's meteoric rise were
common. In 1911, taking stock of his nation's recent development, Horacio
Rivarola, a young constitutional scholar and later rector of the University of
Buenos Aires, listed the criteria against which the course of civilization
should be measured: "statistics, figures, geographical data, history, sociol-

ogy, anthropology, social psychology, education, traditions, political institutions, laws, constitutions—all this is necessary to follow the plot of a nation's development." He concluded "most humbly . . . [that] our institutions have accompanied the moral and material transformations of the Argentine nation." Carlos F. Gómez, a national deputy, said in 1913, "We find ourselves in the presence of a new social organization. Our life has moved enormously away from the past; it is not one of tyranny. . . . The political culture of the population has been raised prodigiously. . . . It is a new country, a new civilization." Yet "despite the advances," Gómez admitted, "we are in reality a country in formation."[3]

The Argentina of 1915 was nearly unrecognizable from the sleepy outpost of thirty years earlier; it was now a hub of world trade and home to nearly 1.5 million people, with a tenfold population increase since 1869.[4] José María Monner Sans, a popular Argentine literary critic, echoed the sense of triumphant social evolution toward civilization. To an audience at the Hispanic American Athenaeum of Buenos Aires, speaking on the sweeping but timely question of "the social function of our generation," he asserted that, thanks to the "advances of the nineteenth century and [its] untidy strength . . . [Argentina] is moving with frightening speed from childhood to adolescence."[5] The political landscape now shifted significantly as well; the middle-class Radical Party, which had long sought to unseat the oligarchy from its monopoly of electoral office, in 1916 won the first popular election and the presidency.

The state's leaders, for all their prideful boasting of Argentina's achievements of the past thirty years, knew that their hard-won and newfound status was tenuous, and they feared that "civilization" could be eroded by tensions boiling not too far under the surface. Modernity's conquerors were not all benevolent. With the war, trade was temporarily halted, and the flow of foreign capital slowed. Economic growth did not repeat its astronomical pre-1914 rates. Recurring economic depressions shook the nation's confidence, and in response to the resulting unemployment and bankruptcies came big strikes and disruptive student demonstrations. Police faced off against these demonstrators, sometimes with the support of right-wing, anti-Semitic, and xenophobic vigilante groups. Social disorder and crime remained constant problems. In 1914 respected criminologist Miguel A. Lancelotti, an associate at the Criminology Institute of Buenos Aires, was inspired to write, "It may seem paradoxical but I am certain that despite the advances of civilization and the new avenues of work, criminality continues to increase in all lands, as does reincidence . . . . Buenos Aires has not escaped this sad honor."[6]

Were crime, degeneracy, racial pollution, and challenges to the social hierarchy really threats to modern Argentina? Or did the state claim to seek to eliminate them to mask its principal goal, the control and discipline of dissidents and the poor? Officials cynically used the language and techniques of science to address social conflict, but their central purpose was to maintain the economic and social status quo and their own power. Thus they saw as the main task of the police to address the nation's "social, political, and economic instability." While disease and poverty could be studied and treated as crimes, the word "crime" itself was frequently a euphemism for challenges to the oligarchy's vision of progress or, in the words of Julio A. Roca, president of the republic for much of the 1880s and 1890s, to "order and administration."[7] Reformers and scientists in Buenos Aires, many of whom had a genuine interest in helping the poor and deranged, did bitterly lament the lack of decent hospitals and funds for welfare and inspection. Yet, the state itself and its principal scientific allies did not address the underlying causes of Argentina's social conflict—poverty, malnutrition, and the exploitation of labor—and focused instead on threats to economic progress and entrenched hierarchy. Anarchists, in particular, were dealt with swiftly and punitively by enhanced police power and repressive laws.

The language of science and medicine proved a useful way to couch the state's efforts to maintain control. Public health, criminology, criminalistics, and penology, more than other disciplines, served to mask larger meanings about social roles and expectations. Rhetoric of cleansing, sanitation, and purity pervaded Argentine political culture and shaped state formation and practices through progressive and regressive regimes. Scientific rationales for punitive state technologies gave expression to a virulent strain of nativism and patriotic obsession. Each successive government sought to purge unfit components from the national body: degenerates, the physically weak, and the dangerous elements of the immigrant masses. Each created institutions of marginalization and stigmatized those who dissented. Behind the government's rationale were pseudoscientific stereotypes that sought to define and identify who was normal and who was dangerous. These emphases explain why Argentines so readily adapted the Italian school of criminal anthropology, along with French degeneration theory, both of which found pathology in the essence or physical traits of the person.

The application of scientific ideas and social engineering in the liberal period, cynical or sincere or both, did not bring an end to the nation's social tensions, or even the effective suppression of disorder, but rather it laid the groundwork, if inadvertently, for exacerbating social conflicts. In the twentieth century the battle over Argentine definitions of civilization would resurface, through military regimes, Peronism, and a bloody "process of national reorganization" by a totalitarian government, against what each regime would consider challenges to Argentine civilization. By appealing to the defense of civilization, Argentina had, in fact, turned its definitions upside down and created a new form of barbarism, one typified by increased state surveillance, by brutal discipline and control, and by a racialized stigmatization of unfavorable and resisting groups. What state officials actually feared was the full participation of their population in Argentine society. The civilization made possible by their practices was superficial at best, and corrupted at worst.

In the quest to maximize Argentina's chances for development on north Atlantic models, the nation's elite focused on the makeup of its citizenry. While Argentine elites had few images before 1900 of who they were, they did have a clear idea of their antiheroes: Indians, blacks, cowboys, immigrants, and workers with a radical political agenda. Argentina had not been the large-scale, highly organized civilization before the European conquest that most other Spanish colonies were. Of the multiple native groups from the Andes to Tierra del Fuego, none had a complex culture like the Aztecs or Incas. Argentina had no such ancient traditions to give it pride and a shared sense of identity at the turn of the twentieth century. The country was ashamed of and, its leaders believed, weakened by its native groups. The now notorious "conquest of the desert" of 1879, a series of struggles over the land typical of earlier New World communities, culminated in a brief and bloody campaign to exterminate or drive out Pampas Indians. In the same vein, after 1900 political discourse on civilization versus barbarism took on a racist perspective. Elites assembled a view of the outsider as pathological, contagious, and dangerous to national stability. They conjured up enemies of the state out of the mass of foreign immigrants who demanded improved working and living conditions and access to citizenship.

Not all scientists had the same perceptions of Argentina's problems, but the state's legal, educational, and penal policies and practices were rationalized by scientific ideas of biology, heredity, blood, degeneration, and psychology. Ideas led to practices. Fingerprinting and other identification techniques helped the state impose boundaries of citizenship and nationality. The scientist-elites' methodology was technical, statistical, administrative, and institutional, with the accoutrements of scientific objectivity. But its premises were not scientifically verifiable, and its outcomes were not always socially beneficial.

Contrary to the leadership's claims, the emerging social "science" did not prevent or cure crime, suffering, or social problems. For most Argentines, poverty and social hierarchy, the stigmatization of immigrant groups and reinforcement of inequity in political participation and civil rights became more rigid in this period. The new ideas about how to identify the dangerous "other" shaped the application of disciplinary techniques and further divided the nation. The unprecedented funds spent on "scientific" measurement of criminals' bodies, on showcase prisons, on fingerprinting, on "experts" in the criminal justice system, and on rewrites of the legal codes did nothing to redress the deep disparities in economic and political privileges that were a genuine plague on the country.

Argentina's "golden era" of economic growth, the period of the "great leap" from 1880 to 1914, thus can be understood best not as a period of great promise that, as historians have tended to believe, was betrayed by external conditions, such as neocolonial terms of trade, or by internal missteps, such as an overreliance on foreign economies and corruption, but rather as an anomalous period in Argentine economic and social history. The nation's prosperity was not as deep as contemporary observers described, nor was wealth distributed more equitably among the people. Long-term prosperity in the form of a large, well-educated, and well-fed middle class did not occur. A middle class, larger than in most other Latin American countries, emerged, but then declined. Rather, at the turn of the century daily life for most of the nation's inhabitants was characterized by poverty and disease. Fluctuating employment rates and huge disparities in income were defining features of the period.[8] Despite the evidence of economic growth, the majority of the population could not afford to live well; as workers' newspapers pointed out, prices were high and wages did not keep up. Studies of nutrition show that the health of most Argentines did not improve during the economic boom.[9]

In spite of the many comparisons offered of Argentina and the United States and other lands of recent settlement at the turn of the century, many significant differences existed. Most important, the United States was 200 years from independence and had achieved stability. Argentina, closer to its own colonial past, was in 1880 a mere fifty years from independence. Those five preceding decades were marked not by democracy and stability but by dictatorship and civil war. Viewed in the long term, the Argentine economy was closer to the rest of a more volatile and impoverished Latin America than it was to that of the wealthy United States, despite the hopeful thinking of Argentina's planners. Even during the peak years of its wealth, underlying patterns ensured that short-term wealth would wreak havoc in the long term. Landownership patterns from the colonial years concentrated wealth in the hands of the few, and a dependent monocrop export economy underlay the expansion. That boom was not directed toward investment in long-term growth strategies but oriented instead toward maximizing the profits of a small elite. Argentina never achieved full economic self-determination; it remained dependent on foreign capital and markets.

The dramatic national wealth that resulted from the late nineteenth-century agricultural export economy did not significantly alter Argentina's political culture. Political elites both at the national and the local levels resorted, when they felt threatened, to violence and political strong-arm tactics. Liberals at the turn of the century did use language and employed practices considered acceptable to modern "civilized" societies and aimed to avoid the overt brutality of the colonial and mid-nineteenth-century regimes. Yet, they called on the army ever more frequently to quell social unrest and other challenges to state power, including the many local uprisings in the provinces. As the military grew in strength and authority, it emerged as a key player in politics; the army was seen as patriotic in a pure sense, as adhering to "traditional" Argentine culture without the corruption and scandals that made political figures suspect. As a result, the officer corps of the military took on more political roles.[10] Compulsion in the modern era was thus masked by the language of progress.

The influential Generation of 1880, the scientists and bureaucrats of Argentina's most prosperous period, profoundly shaped the state's ability to foster productive modernization as well as to deploy more efficient means of social control. The contradictions between Argentine liberalism and state violence erupted in 1919, when, during a bloody Buenos Aires summer, civilians, police officers, and national troops joined hands to counter what they saw as a dangerous upwelling of worker violence. This episode, called the *semana trágica* or "tragic week," coincided with the emergence of the Argentine Patriotic League, a virulently racist and right-wing vigilante group closely aligned with the Buenos Aires police force and the military. In violent street battles, these freelance and governmental forces suppressed workers and attacked Jewish immigrant neighborhoods on the grounds that they were harboring strikers. The irony was that such repression occurred not under military or authoritarian rule but under a liberal, progressive, modern government at the zenith of Argentina's most flourishing era. The moderate Unión Civica Radical, known as the Radical Party, which consisted largely of middle-class reformers, had swept into power in 1916 on a wave of popular votes made possible by the electoral reform in 1912, capping a decades-long struggle against the entrenched, elitist oligarchy. Radical leaders, including the new president of the republic, Hipólito Yrigoyen, were thus both the successors and the amenders of a long line of liberal thought in Argentina—subscribers, as its leaders understood liberalism, to notions of progress and modernization.[11]

Under a facade of order, the government was in crisis. By the 1920s, violent suppression of public demonstrations and popular political protest had come to seem, even to the reformist Radicals, necessary to modern governance, social order, and ultimately, the peace and prosperity of the nation. The Radicals proved just as, or even more, willing to use force to maintain order than the oligarchy they had electorally replaced had been. Xenophobic and racist theories, which had developed under the Generation of 1880's leadership, also grew and flourished on the Radicals' watch. Organizations that coalesced in the name of science, such as the Argentine Eugenics Society and the Argentine Association of Biotypology, Eugenics, and Social Medicine, were firmly aligned with the political right wing. While some proponents of turn-of-the-century racial science, convinced of concepts of group superiority and inferiority, sought social reforms along with more punitive measures, those who inherited the existing groups or who founded similar institutions after the 1910s supported racial hierarchy, segregation, and eugenic immigration measures. In 1941 Arturo R. Rossi, the president of the Association of Biotypology, Eugenics, and Social Medicine, described the organization's goal as "defending white civilization . . . with energy and tenacity" against the "profound polymorphism of our people." He defended this goal in the name of science.[12]

In 1930 Yrigoyen, the Radical president, was ousted in a military coup, ending fifty years of elected governments. A series of military governments followed. In 1946, on taking office as president, the populist leader Juan Perón targeted the oligarchic elite as betrayers of the nation. Perón had emerged from a military culture and in 1943, as director of the National Labor Office under the ruling regime, used his post as a springboard to monolithic power. With offers of generous social benefits and official recognition, Perón craftily co-opted the nation's numerous industrial workers and won their loyalty.

Perón had gained enough support from elites and the middle class on the promise of keeping the plebian mobs in order, though most of the entrenched Eurocentric upper class despised him for policies that favored the nation's new industrialist class and for his vulgar lifestyle. His formidable second wife Eva, a former B-movie star, was in the view of the old elite a harlot, and Perón's embrace of the "shirtless ones" and the "little black heads" was considered low class and repugnant. (Perón, too, emulated European culture, but in his case, he embraced the punitive and authoritarian aspects of fascist Europe.)[13] His strong-arm tactics appealed to those who thought the nation was in constant risk of disruption by a mobilized working class, still deemed "foreign."[14] Like his liberal predecessors, Perón focused on the concept of Argentine civilization and employed the rhetoric of progress and civilization to vilify his enemies. He, too, exploited medical metaphors to justify repressive practices, and the Ramos Mejía Hospital—named for a prominent member of the Generation of 1880, psychiatrist José María Ramos Mejía—became a central location for police torture of his political opponents.[15] Perón was overthrown in 1955 by the same means his group had employed to take power—a military coup—and though for the next two decades he lived in exile, he returned to power in 1973. He died the following year, but Peronism as a political movement continued to hold sway over many Argentines.

In the 1970s military rulers again cynically manipulated the concept of civilization, also bolstered with medical metaphors. In a 1976 coup, a ruthless military junta initiated an eight-year nightmare for Argentina; 30,000 people would disappear, most of whom were never found, dead or alive. (In the 1990s former military officials confessed that they had supervised the burial of these people in unmarked graves or dumped their bodies from helicopters into the sea.) Women prisoners were raped and sexually tortured. Infants and young children of the disappeared were stolen and adopted by military families. The goal of the "*proceso*" (process), as the military rulers referred to their campaign against their own people, was the defense and renewal of, in the words of a 1979 government document, "Western, Christian civilization." Overtly racist and anti-Semitic, the military regime frequently and openly promoted the idea of Jewish conspiracy and justified a racial cleansing of their society in defense of true Argentine civilization. The language of illness, infection, and sanitation permeated the regime's justifications for terror against the population. The mothers of the disappeared—who gathered to demand information about their children's fates—were branded *locas*, madwomen. Leftists of any gender or economic class were identified as germs and unhealthy "others" to be eliminated completely. Rear Admiral César A. Guzetti wrote in the newspaper *La Opinión* in 1976 that "the social body of the country is contaminated by an illness that in corroding its entrails produces antibodies. These antibodies must not be considered in the same way as [the original] microbe. As the government controls and destroys the guerrilla, the action of the antibody will disappear. . . . This is just the natural reaction of a sick body."[16]

Events of the early 1980s, including the state terror and lack of economic recovery, would cause the military to lose legitimacy. A devastating defeat by the British in the brief 1982 Malvinas War (known to English speakers as the Falkland Islands) would bring the military down, and in 1984 the Radical Party's Raúl Alfonsin would win in a landslide, the first democratically elected president since Perón. Alfonsin would be followed in 1992 by Carlos Menem, the son of Syrian immigrants and a champion of extreme neoliberal economic policies, including wholesale privatization of state companies, layoffs of state employees, and drastic cuts in social spending. The memory of terror remained woven into the fabric of everyday life. Despite a public outcry, President Menem pardoned most of the military officers who had overseen the torture and disappearance of thousands, even though he had been jailed himself. In the 1990s two bombs were detonated in Buenos Aires against Jewish targets, and a journalist investigating police corruption, José Luis Cabezas, disappeared; both actions were linked to the notoriously corrupt, right-wing provincial police of Buenos Aires.[17] By 2000, Argentina had acquired institutional democracy, but worsening social inequality remained, and, as with many other Latin American countries, the threat of authoritarianism had not disappeared.

Ironically, both the rise and then the decline of Argentina's promise were the result not of elite self-interest or economic or social backwardness but, disturbingly, of the liberal scientific impulse to "cure" social ills. On these grounds and in the antihumanist and patronizing form of social progress devised to contain and control inferior groups were the real seeds of national decay. A powerful alliance between medicine, science, and state power built this self-destructive modern Argentine political culture. A coalition of state officials and scientists that called itself "progressive," through exploitation of science and use of punitive controls, turned Argentina's most hopeful moment into tragedy with repercussions well into the twentieth century.

When the state and science collude, a contradiction is implicit and ex-plicit in that relationship. Argentine scientists who considered themselves neutral, reporting without bias the real facts of society and human nature, seemed to be unaware that in social science compiling and evaluating evi-dence may be tainted by human perspectives. In measuring skulls and as-suming negative racial and gender characteristics, these scientists worked from deep premises that shaped their selection and evaluation of evidence. When, in turn, their oligarchic patrons found their findings useful, their science was elevated and, if inadvertently, used for reasons they could not control. Science, to the liberal modernizers of the Generation of 1880, had meant all good things—efficiency, health, purity, cleanliness, order—but it had other, less overt intentions and unintended outcomes, including the control of presumed-to-be lesser people. The state's priority was not really prevention of crime, as it claimed, but preservation of the economic and social hierarchy on which the export economy rested. Social order, achieved by stigmatization and control, was critical to this goal. Science, innocently or otherwise, provided a modern, progressive ideology and language to support that goal.

State exploitation of science did have its detractors. Critics from the left, including anarchists, socialists, and members of other popular movements, publications, and public demonstrations, mocked the government for in-eptitude and pomposity.[18] Traditionalists from the right harshly criticized the state for turning its back on authentic Argentine values. Resistance emerged, too, from the mass of society, as individuals refused to subject themselves to medical exams or fought against fingerprinting. But though some scientists were critical of the oligarchic status quo and sympathetic to the plight of workers and the poor and, to a certain extent, women, the scientific ideas taken up and implemented by those in power were the ideas that justified government's own interests. The oligarchy had found the explanations that blamed the poor for social turmoil and for their own misery the most compelling. The outcome was that surveillance increased while living standards did not.

It was democracy that the oligarchy feared. Turn-of-the-century elites did not trust the Argentine masses, whether native, foreign, mixed-race, or provincial creoles, believing them incapable of carrying out democracy. Mass mobilizations for improved housing, nutrition, wages, and political representation were the specter elites fought against. Their urge for order, combined with hubris, justified their elitist, top-down governance. Moder-ate reformist legislation and the eventual male franchise of 1912 with secret balloting would coexist with occasions of federal intervention in the provinces and continued harsh repression of labor activism.[19] Women were denied the vote until midcentury. When it came to opening venues for wider political participation, Argentina's elite fell back on the entrenched habits of their class: marginalization of the masses and control and repression of their opponents. The only organized force that emerged after 1930 with the potential to unseat this entrenched autocracy's hold on power was Peronism. Though adversarial to its predecessors, it resembled in certain respects the preceding regimes. It, too, relied on violence, exclusion, and the repression and scapegoating of those believed to defy it, though not in the name of science, reason, and progress.

Argentina's eventual policy direction toward the authoritarianism of the mid-twentieth century was not a puzzle but the direct consequence of an unresolved earlier paradox between progress and repression.[20] The rudiments of later reactionary state violence were rationalized, strengthened, and mobilized during Argentina's rare period of liberal democracy. Liberal efficiency, at its heart a partnership between science and the state, had played an essentially reactionary role. Together they, in the name of modernization and progress, helped shape the trend toward an antidemocratic authoritarianism. Thus, Argentine history shows not a break but continuity; there are fundamental links between the pre–World War I period, generally understood as essentially liberal and democratic, and Peronism and the other protofascist governments that came after, especially in terms of elite responses to social tensions and the state capacity for modern, rationalized social control. Argentine liberalism in its foundational period laid the groundwork for violation of its own inherent principles.

The rhetoric of science, universal truth, and national progress, acquired in Argentina's "golden age," did not alleviate the unresolved historical tensions between democracy and populism, on the one hand, and order and repression on the other. The lasting legacy of scientific innovation in turn-of-the-century Argentina would indeed lead to a new form of barbarism. As the twentieth century progressed, Argentina had yet to resolve its most urgent dilemma—how to create a civilized nation.

Notes

Introduction

1. Larry Rohter, "A Street Battle Rages in Argentina's 150 Years War," *New York Times*, 14 August 2003.

2. Nicolas Shumway, *The Invention of Argentina: History of an Idea* (Berkeley: University of California Press, 1991), x.

3. See Marguerite Feitlowitz, *A Lexicon of Terror: Argentina and the Legacies of Torture* (New York: Oxford University Press, 1998).

4. Karl Monsma, "Beyond Dependency: Historical Sociology and Social Change in the Southern Cone of South America," *Comparative Studies in Society and History* 33, no. 4 (1991): 795. See also Jorge Schvarzer, "The Argentine Riddle in Historical Perspective," *Latin American Research Review* 27, no. 1 (1992): 169–81.

5. Carlos Waisman, *The Reversal of Development in Argentina: Postwar Counter-*

revolutionary Policies and Their Structural Consequences (Princeton, N.J.: Princeton University Press, 1987).

6. See, for instance, Mark Falcoff and Ronald H. Dolkart, *Prologue to Perón: Argentina in Depression and War, 1930–1943* (Berkeley: University of California Press, 1975). Waisman, in his influential 1987 book, *Reversal of Development in Argentina*, xi, also places this reversal in the 1930s and 1940s.

7. On the paradoxes of turn-of-the-century liberal reforms in Argentina, see Ricardo Salvatore, "Death and Liberalism: Capital Punishment after the Fall of Rosas," in *Crime and Punishment in Latin America: Law and Society Since Late Colonial Times*, ed. Ricardo Salvatore, Carlos Aguirre, and Gilbert Joseph (Durham, N.C.: Duke University Press, 2001), 308–41. On women and liberalism see Kristin Ruggiero, "Honor, Maternity, and the Disciplining of Women: Infanticide in Late Nineteenth-Century Buenos Aires," *Hispanic American Historical Review* 72, no. 3 (1992): 353–73; and Lila M. Caimari, "Whose Prisoners Are These? Church, State, and *Patronatos* and the Rehabilitation of Female Criminals (Buenos Aires, 1890–1970)," *Americas* 54, no. 2 (1997): 185–208.

8. On the oligarchy's strange mixture of progressive liberal reform and social and political conservatism, see Natalio Botana, *El orden conservador. La política argentina entre 1880 y 1916* (Buenos Aires: Sudamerica, 1994).

9. Alfredo G. Kohn-Loncarica, "Ciencia y estado en la Argentina. Una perspectiva histórica de sus relaciones," *Revista de Propuesta y Control* 16, no. 22 (1992): 2475–97.

10. See Kristin Ruggiero, *Modernity in the Flesh: Medicine, Law, and Society in Turn-of-the-Century Argentina* (Stanford, Calif.: Stanford University Press, 2004); Nancy Leys Stepan, *"The Hour of Eugenics": Race, Gender, and Nation in Latin America* (Ithaca, N.Y.: Cornell University Press, 1991); Eduardo Zimmermann, *Los liberales reformistas. La cuestión social en la Argentina, 1890–1916* (Buenos Aires: Editorial Sudamericana, 1994); and Donna Guy, *Sex and Danger in Buenos Aires: Prostitution, Family, and Nation in Argentina* (Lincoln: University of Nebraska Press, 1990).

Chapter One

1. See Kristine L. Jones, "Comparative Raiding Economies," in *Contested Ground: Comparative Frontiers on the Northern and Southern Edges of the Spanish Empire*, ed. Donna J. Guy and Thomas E. Sheridan (Tucson: University of Arizona Press, 1998), 97–114.

2. Afro-Argentines, despite their diminishing numbers after 1850, retained a distinct presence in Buenos Aires. George Reid Andrews, *The Afro-Argentines of Buenos Aires, 1800–1900* (Madison: University of Wisconsin Press, 1980), 4.

3. Nicholas Canny and Anthony Pagden, eds., *Colonial Identity in the Atlantic World, 1500–1800* (Princeton, N.J.: Princeton University Press, 1987), 66.

4. Domingo Faustino Sarmiento, *Civilización i barbarie, vida de Juan Quiroga Fac-*

undo (Santiago, Chile: Imprenta del Progreso, 1845). The book has been recently translated as *Facundo: Civilization and Barbarism*, trans. Kathleen Ross (Berkeley: University of California Press, 2003). The central theme of civilization and barbarism has been picked up in two important scholarly books: Nicolas Shumway, *The Invention of Argentina: History of an Idea* (Berkeley: University of California Press, 1991); and Francine Masiello, *Between Civilization and Barbarism: Women, Nation, and Literary Culture in Modern Argentina* (Lincoln: University of Nebraska Press, 1992).

5. Juan Bautista Alberdi, *Bases y puntos de partida para la organización política de la República Argentina* (1852; Buenos Aires: Ediciones Jackson, 1938), 66–69. Unless otherwise noted, all translations from the Spanish and French are the author's.

6. Tulio Halperín Donghi, "Argentine Counterpoint: Rise of the Nation, Rise of the State," in *Beyond Imagined Communities: Reading and Writing the Nation in Nineteenth-Century Latin America*, ed. Sara Casto-Klarén and John Charles Chasteen (Washington, D.C.: Woodrow Wilson Center Press, 2003), 33–53.

7. David Rock, *Argentina, 1516–1987* (Berkeley: University of California Press, 1987), 124, 132, 153; James R. Scobie, *Buenos Aires: Plaza to Suburb, 1870–1910* (New York: Oxford University Press, 1974), 97.

8. Rock, *Argentina*, 132, 172; Scobie, *Buenos Aires*, 129–35.

9. Juan A. Alsina, *La inmigración Europea en la República Argentina* (Buenos Aires, 1898), 331. Emphasis in original.

10. Santiago Vaca-Guzmán, *La naturalización de los extranjeros* (Buenos Aires: Casa Editoria Jacobo Peuser, 1891), 85.

11. Nicolás Avellaneda, "Introduction," in Luis Jorge Fontana, *El gran Chaco* (1881; Buenos Aires: Solar Hachette, 1977), 27, 39.

12. Fontana, *El gran Chaco*, 136–37.

13. The phrase "*conquista del desierto*" has been translated both as "conquest of the desert" and "conquest of the wilderness." The area in question, in the view of the military commanders, was a desert not in the topographical sense, but in the cultural sense.

14. Carlos Bunge, *Nuestra América. Ensayo de psicología social* (1903; Buenos Aires: Rosso y Compañia, 1918), 230. Emphasis in original.

15. Ibid., 49.

16. Lucas Ayarragaray, *La anarquía argentina y el caudillismo. Estudio psicológico de los origins nacionales, hasta el año xxix* (Buenos Aires: Lajouane, 1904), 2, 15, vi;

17. Ibid., 130–31, 132.

18. José Ingenieros, "La anarquía argentina y el caudillismo," *APCCA* 3 (1904): 519, 526; Lucas Ayarragaray, "La constitución étnica argentina y sus problemas," *APCCA* 11 (1912).

19. Bunge, *Nuestra América*, 49, 168, 169, 212. Emphasis in original.

20. Victor R. Pesenti, "La influencia de la civilización sobre el movimiento de la criminalidad" (Law thesis, UBA, 1901), 113.

21. Bunge, *Nuestra América*, 51.

22. Translated and cited in David Rock, *State Building and Political Movements in Argentina, 1860–1916* (Stanford, Calif.: Stanford University Press, 2002), 216.

23. Per capita rates from Carlos Waisman, *The Reversal of Development in Argentina: Postwar Counterrevolutionary Policies and Their Structural Consequences* (Princeton, N.J.: Princeton University Press, 1987), 5; Rock, *Argentina*, 172. For a typical, optimistic assessment of Argentina's late nineteenth-century modernization efforts, see Reginald Lloyd, ed., *Impresiones de la República Argentina en el siglo viente* (London: Lloyd's Publishing Company, 1911).

24. On the terms "liberal" and "conservative" in Argentine politics of the nineteenth century, see Rock, *State Building*, 3–4, 9–10.

25. Francisco de Veyga, "Anarquismo y anarquistas. Estudio de antropología criminal," *ADNH* 7, no. 20 (1897): 451.

26. Scobie, *Buenos Aires*, 235–38.

27. Rock, *Argentina*, 141.

28. Bunge, *Nuestra América*, 212.

29. José C. Moya, *Cousins and Strangers: Spanish Immigrants in Buenos Aires, 1850–1930* (Berkeley: University of California Press, 1998), chap. 2; Carl Solberg, *Immigration and Nationalism: Argentina and Chile, 1890–1914* (Austin: University of Texas Press, 1970), 33–38. For official immigrant statistics (including nationality) see Alberto Kleiner, ed., *Publicidad ofical sobre Argentina como país de inmigración* (1903; Buenos Aires: Poligono SRL, 1983), 30.

30. On racial ideas in Latin America, see Nancy Appelbaum, Anne S. Macpherson, and Karin Alejandra Rosemblatt, eds., *Race and Nation in Modern Latin America* (Chapel Hill: University of North Carolina Press, 2003); and Richard Graham, ed., *The Idea of Race in Latin America, 1870–1940* (Austin: University of Texas Press, 1990).

31. Eugenio Cambaceres, *En la sangre* (1887; Buenos Aires: Ediciones Colihue, 1980), 76.

32. Nancy Leys Stepan, *"The Hour of Eugenics": Race, Gender, and Nation in Latin America* (Ithaca, N.Y.: Cornell University Press, 1991), 142–43. On anti-Semitism in Argentina, see Sandra McGee Deutsch, "The Argentine Right and the Jews, 1919–1933," *Journal of Latin American Studies* 18, no. 1 (1986): 113–34.

33. Donghi, "Argentine Counterpoint," 52.

34. Wilde quoted in José Luis Romero, *A History of Argentine Political Thought* (Stanford, Calif.: Stanford University Press, 1963), 185.

35. For a general description, see Scobie, *Buenos Aires*, 146–52. On the utopian spatial visions of modernizing Buenos Aires, see Jorge F. Liernur and Graciela Silvestri, *El umbral de la metrópolis. Transformaciones técnicas y cultura en la modernización de Buenos Aires (1870–1930)* (Buenos Aires: Sudamericana, 1993).

36. Tulio Halperín Donghi, *Historia de la Universidad de Buenos Aires* (Buenos Aires: Editorial Universitaría de Buenos Aires, 1962).

37. Pesenti, "La influencia de la civilización."

Chapter Two

1. Oscar Terán, *José Ingenieros. Pensar la nación* (Buenos Aires: Alianza Editorial, 1986), 35.

2. For studies revising the idea of Latin American science as peripheral, see Julia Rodriguez, "South Atlantic Crossings: Fingerprints, Science, and the State in Turn-of-the-Century Argentina," *American Historical Review* 109, no. 2 (2004): 387–416; Marcos Cueto, *Excelencia científica en la periferia* (Lima, Peru: Grupo de Analisis para el Desarrollo, 1989); Marcos Cueto, "Laboratory Styles in Argentine Physiology," *Isis* 86 (1995): 228–46; and Nancy Leys Stepan, *"The Hour of Eugenics": Race, Gender, and Nation in Latin America* (Ithaca, N.Y.: Cornell University Press, 1991).

3. Anastasio Alfaro, "Los infanticidios en la época colonial," *APCCA* 2 (1903): 48.

4. Eugenio Cambaceres, *Sin rumbo* (1885; Madrid, Spain: Anaya, 1971); Eugenio Cambaceres, *En la sangre* (1887; Buenos Aires: Ediciones Colihue, 1980).

5. Piñero quoted in Hugo Vezzetti, *El nacimiento de la psicología en la Argentina* (Buenos Aires: Puntosur, 1987), 44.

6. Francis Galton to Juan Vucetich, 11 January 1897, miscellaneous folder, AR.

7. Benigno Trigo, *Subjects of Crisis: Race and Gender as Disease in Latin America* (Hanover, N.H.: University Press of New England, 2000), chap. 4.

8. Hugo E. Biagini, ed., *El movimiento positivista argentino* (Buenos Aires: Belgrano, 1985); José Luis Romero, *A History of Argentine Political Thought* (Stanford, Calif.: Stanford University Press, 1963); Ricaurte Soler, *El positivismo argentino* (Buenos Aires: Paidos, 1968); Oscar Terán, *Positivismo y nación en la Argentina* (Buenos Aires: Puntosur, 1987).

9. Thomas Glick, *Darwin y el darwinismo en el Uruguay y en América Latina* (Montevideo: Universidad de la República, 1989), 35–56.

10. "El campo de las calaveras en la isla de Luzon (Filipinas)," *SM*, 28 April 1898, cclxxviii–cclxxx. Reprinted from the *Cronica de Ciencias Médicas de Filipinas*.

11. Eduardo A. Zimmermann, "Racial Ideas and Racial Reform: Argentina, 1890–1916," *Hispanic American Historical Review* 72, no. 1 (1992): 23–46.

12. Nancy Leys Stepan, "Biological Degeneration: Races and Proper Places," in *Degeneration: The Dark Side of Progress*, ed. J. Edward Chamberlin and Sander L. Gilman (New York: Columbia University Press, 1985), 97–120; Eric T. Carlson, "Medicine and Degeneration: Theory and Praxis," in *Degeneration*, ed. Chamberlin and Gilman, 121–44; Daniel Pick, *Faces of Degeneration: A European Disorder* (Cambridge, U.K.: Cambridge University Press, 1989).

13. Stepan, *"Hour of Eugenics,"* chap. 2.

14. Luis María Drago, *Los hombres de presa* (Buenos Aires: Lajouane, 1888), 22–23.

15. Ruth Harris, *Murders and Madness: Medicine, Law, and Society in the Fin de Siècle* (Oxford, U.K.: Oxford University Press, 1989); Marie-Christine Leps, *Apprehending the Criminal: The Production of Deviance in Nineteenth-Century Discourse* (Durham, N.C.: Duke University Press, 1992); Robert Nye, *Crime, Madness,*

and Politics in Modern France: The Medical Concept of National Decline (Princeton, N.J.: Princeton University Press, 1984); Pick, *Faces of Degeneration*; Ann-Louise Shapiro, *Breaking the Codes: Female Criminality in Fin-de-Siècle Paris* (Stanford, Calif.: Stanford University Press, 1996). On the history of criminology and penology in Latin America, see Lyman L. Johnson, ed., *The Problem of Order in Changing Societies: Essays on Crime and Policing in Argentina and Uruguay, 1750–1940* (Albuquerque: University of New Mexico Press, 1990); Ricardo D. Salvatore and Carlos Aguirre, eds., *The Birth of the Penitentiary in Latin America: Essays on Criminology, Prison Reform, and Social Control, 1830–1940* (Austin: University of Texas Press, 1996); and Ricardo Salvatore, Carlos Aguirre, and Gilbert Joseph, eds., *Crime and Punishment in Latin America: Law and Society Since Late Colonial Times* (Durham, N.C.: Duke University Press, 2001).

16. Mary Gibson, *Born to Crime: Cesare Lombroso and the Origins of Biological Criminology* (Westport, Conn.: Praeger, 2002); Pick, *Faces of Degeneration*.

17. On Lombroso's view of women, see Gibson, *Born to Crime*, chap. 2.

18. From Savitch's "Introduction" in Gina Lombroso-Ferrero, *Criminal Man, According to the Classification of Cesare Lombroso* (New York: G. P. Putnam's Sons, 1911), xi.

19. Hermann Mannheim, ed., *Pioneers in Criminology* (Montclair, N.J.: Patterson Smith, 1972), 250.

20. José Ingenieros, "Etiología y terapeutica del delito," *SM*, 10 August 1899, 282–84.

21. In Argentina the critics of modernity opposed the biological determinism of positivist criminology on religious grounds. After 1900, a secular critique criticized positivism as materialist and narrowly scientistic. On the antipositivists in Argentina, see David Rock, "Intellectual Precursors of Conservative Nationalism in Argentina, 1900–1927," *Hispanic American Historical Review* 67, no. 2 (1987): 271–300; and David Rock, *Authoritarian Argentina: The Nationalist Movement, Its History, and Its Impact* (Berkeley: University of California Press, 1993), 39.

22. "Antropología y craneología," *SM*, 11 August 1898, dxiv–dxxiii; H. A. Kozloff, "Los niños de padres criminales y no criminales," *SM*, 16 February 1899, 63–64.

23. This 1876 article was entitled, "La fisica en la filosofía"; cited in Hebe Clementi, "José María Ramos Mejía," in *El movimiento positivista argentino*, ed. Biagini, 391.

24. Cornelio Moyano Gacitúa, *Curso de ciencia criminal y derecho penal argentino* (Buenos Aires: Lajouane, 1899).

25. F. S. L. Lyons, *Internationalism in Europe, 1815–1914* (Leyden: A. W. Sythoff, 1963).

26. Emilio R. Coni, *Progrès de l'hygiène dans la République Argentine* (Paris: Librairie J-B. Bailliere et Fils, 1887).

27. Cited in Francisco Durá, *Naturalización y expulsión de extranjeros* (Buenos Aires: Coni Hermanos, 1911), 63. See also "République Argentine," in *Le Congrès Pénitentiaire International de Stockholm: Mémoires et Rapports* (Stockholm: Bureau de la Commission Pénitentiaire International, 1879).

28. *Primera reunión del Congreso Científico Latino Americano celebrada en Buenos Aires del 10 al 20 de abril de 1898* (Buenos Aires: Compañia Sudamericana de Billetes de Banco, 1898).

29. Piñero quoted in Vezzetti, *El nacimiento de la psicología*, 44.

30. José Ingenieros, "Al partir," in *Crónicas de viaje* (Buenos Aires: L. J. Rosso, 1919), 253–54.

31. Francisco Ramos Mejía, *Principios fundamentales de la escuela positiva de derecho penal* (Buenos Aires: Est. Tip. De El Censor, 1888), 3, 11–12, 18, 17, 21.

32. José María Ramos Mejía, "Apuntes clinicos sobre traumatismo cerebral" (Medical thesis, UBA, 1879); José María Ramos Mejía, *Las neurosis de los hombres célebres en la historia argentina* (Buenos Aires: M. Biedma, 1878).

33. Francisco de Veyga, *Anarquismo y anarquistas. Estudio de antropología criminal* (Buenos Aires, 1897); see also Francisco de Veyga, "Anarquismo y anarquistas. Estudio de antropología criminal," *ADNH* 7, no. 20 (1897): 437–55.

34. Abelardo Levaggi, *Historia del derecho penal argentino* (Buenos Aires: Perrot, 1978), 210.

35. Quoted in Rosa del Olmo, *América Latina y su criminología* (Mexico City: Siglo Veintiuno, 1987), 135.

36. Ramos Mejía, *Principios fundamentales*, 43.

37. Drago, *Los hombres de presa*, 138.

38. Ricardo Gonzales Leandri, "La profesion médica en Buenos Aires: 1852–1870," in *Política, médicos, y enfermedades. Lecturas de historia de la salud en la Argentina*, ed. Mirta Zaida Lobato (Buenos Aires: Editorial Biblos, 1996), 21–53. See also "Documentos oficiales," *RMQ* 4, no. 15 (8 November 1867): 226–27; and "La Junta de Higiene Pública, " *RMQ* 5, no. 16 (23 November 1867): 242–43.

39. Coni, *Progrès de l'hygiène*, 262–64, 249–50.

40. See Emilio R. Coni, *Código de higiene y medicina legal de la República Argentina para uso de médicos, abogados, farmacéuticos, etc.* (Buenos Aires: Librería de Juan Etchepareborda, 1891), 627.

41. Alfredo Kohn Loncarica and Abel Agüero, "El contexto médico," in *El movimiento positivista argentino*, ed. Biagini, 131–32.

42. José María Ramos Mejía to dean of the medical school, 19 February 1891, personnel folder "José María Ramos Mejía," Archivo de la Facultad de Medicina, UBA.

43. Francisco de Veyga to Dean José F. Baca, 23 October 1899, personnel folder "Francisco de Veyga," Archivo de la Facultad de Medicina, UBA.

44. Emilio R. Coni, *Movimiento de la población de la ciudad de Buenos Aires durante el año 1880* (Buenos Aires: Pablo E. Coni, 1881), contains data on immigration, public health, marriage, infant mortality, welfare, and hospitals.

45. G. Udaondo to Minister of the Interior José V. Zapata, 19 October 1891, *ADNH* 1, no. 10 (1891): 613–14.

46. Series *"Cultura,"* file 52, 13 November 1885, Archivo Historico de la Municipalidad de Buenos Aires. In addition, the *Círculo Médico Argentino* had its own museum with a pathological anatomy section; see *ACMA* 1, no. 1 (20 August 1877): 118–21.

47. Samuel Gache, "Patogenía del suicidio en Buenos Aires," *ACMA* 7 (1883–84): 568.

48. Alfaro, "Los infanticidios en la época colonial," 48.

49. Miguel A. Lancelotti, "La herencia en la criminalidad," *Revista Nacional* 25 (1898): 402.

50. Miguel Cané to the dean, undated [1878], page 18, file 7, "Discurso de la transmisión al decanato en la Facultad de Filosofía y Letras," Archivo Miguel Cané, AGN.

51. Norberto Piñero, "Programa oficial. Curso de derecho penal," cited in Rosa del Olmo, *Criminología argentina. Apuntes para su reconstrucción historica* (Buenos Aires: Depalma, 1972), 4–5.

52. *Facultad de ciencias médicas—Biblioteca. Catálogo de la colección de tésis, 1827–1917* (Buenos Aires: Flaiban, 1918), 442–45; Marcial R. Candioti, *Bibliografía doctoral de la Universidad de Buenos Aires y catálogo cronológico de las tésis en su primer centenario 1821–1920* (Buenos Aires: Ministerio de Agricultura, 1920), 275–315.

53. Some examples include Kozloff, "Los niños de padres criminales y no criminales"; "Instituto de Medicina Legal," *ADNH* (1896): 439–48; Veyga, "Anarquismo y anarquistas"; and Francisco de Veyga, "La locura en la historia. Juicios sobre la obra del Dr. Ramos Mejía," *ADNH* (1897): 334–35.

54. Drago, *Los hombres de presa*.

55. The Italians, interestingly, emphasized the Lombrosian influence on Drago's book by translating the title as *The Born Criminal*. Luis María Drago, *I Criminalinato* (Torino: Fratelli Bocca, 1890).

56. Drago, *Los hombres de presa*, 32, 139–82.

57. Rodolfo Rivarola, *Derecho penal argentino. Tratado general y de la legislación actual comparada con las reformas proyectadas y con legislaciones de lengua española* (Buenos Aires: Rivadavia, 1910).

58. Antonio Dellepiane, *Las causas del delito* (Buenos Aires: Pablo Coni, 1892); Antonio Dellepiane, *El idioma del delito* (Buenos Aires: Arnoldo Moen, 1894).

59. For articles with a national focus in *CM*, see M. Carlés, "Atavismo Pampa," *CM* 1 (1898): 57–58, 84–85, 133–34; M. A. Lancelotti, "Civilización y delito," *CM* 2 (1899): 406–12; V. Arreguine, "El prejuicio patriótico como fuente de delincuencia colectiva," *CM* 2 (1899): 413–17; M. A. Lancelotti, "Curso de ciencia criminal," *CM* 2 (1899): 440–41; M. A. Lancelotti, "El factor económico en la producción del delito," *CM* 3 (1900): 495–500; and José Ingenieros, "Escuela positiva en nuestra enseñanza universitaria," *CM* 3 (1900): 616–18.

60. José Ingenieros, "Nueva clasificación de los delincuentes," *APCCA* 5 (1906): 30–39.

61. Drago, *Los hombres de presa*, 125.

62. "Las instituciones penales en la América de lengua española," *APCCA* 7 (1908): 230.

63. Horacio Areco, "Ciencia, psicología, y derecho," *APCCA* 11 (1912): 565.

64. See, for example, *Boletín mensual de estadística* (La Plata) 5, no. 7 (July 1899).

65. Miguel Lancelotti to Juan Vucetich, 12 September 1900, correspondence folders, AV.

66. José Ingenieros to Juan Vucetich, 11 June 1911, correspondence folders, AV.

67. "La dactiloscopia," *APCCA* 5 (1906): 369–70.

68. Félix Pacheco, "Identificación de los delicuentes. Ventajas del sistema dactiloscópico," *APPCA* 2 (1903): 228.

69. Afranio Peixoto quoted in Luis Reyna Almandos, *Origen del Vucetichismo. Sistema dactiloscópio argentino* (Buenos Aires: Juan Alsina, 1909), 44.

70. Federico Oloríz Aguilera, "Sobre identificación dactiloscópica," *APCCA* 9 (1910): 556.

71. Quoted in Juan Vucetich, "Historia sintética de la identificación," *Revista de Identificación y Ciencas Penales* 7 (1931): 29, 44.

72. José Ingenieros, review of *Delitti vecchi e delitti nuovi* by Cesare Lombroso, *APCCA* 1 (1902): 316.

Chapter Three

1. José Ingenieros, "Imperialismo," in *Crónicas de viaje* (Buenos Aires: L. J. Rosso, 1919), 192, 195, 201.

2. Ingenieros, "Volviendo al terruño," in ibid., 258.

3. Emilio R. Coni, *Progrès de l'hygiène dans la République Argentine* (Paris: Librairie J-B. Bailliere et Fils, 1887). See also Emilio R. Coni, *Movimiento de la población de la ciudad de Buenos Aires durante el año 1880* (Buenos Aires: Pablo E. Coni, 1881), which contains data on immigration, public health, marriage, infant mortality, welfare, and hospitals. For a statistical study linking prostitution with syphilis and illegitimacy see Emilio R. Coni, *Higiene aplicada. La provincia de Corrientes* (Buenos Aires: Pablo Coni, 1898), 343–46.

4. Antonio Dellepiane, *Las causas del delito* (Buenos Aires: Pablo Coni, 1892), 22–23.

5. In his early career Ingenieros used his family's original Italian name "Ingegnieri" or the intermediate "Ingegnieros" until he fully Hispanicized it in 1913. On Ingenieros's life and work see Angel Rodriguez Kauth, *José Ingenieros* (Buenos Aires: Almagesto, 1996); and Oscar Terán, *José Ingenieros. Pensar la nación* (Buenos Aires: Alianza Editorial, 1986).

6. For example, see José Ingenieros, review of Spencer's "Facts and Comments," *APCCA* 1 (1902): 440–41.

7. Mariano Ben Plotkin, "Freud, Politics, and the Porteños: The Reception of Psychoanalysis in Buenos Aires, 1910–1943," *Hispanic American Historical Review* 77, no. 1 (1997): 45–74; Hugo Vezzetti, ed., *Aventuras de Freud en el país de los argentinos. De José Ingenieros a Enrique Pichon Rivière* (Buenos Aires: Paídos, 1996).

8. Ingenieros edited the *Archivos* between 1902 and 1913; during this period it changed names once. It started in 1902 as the *Archivos de Criminología, Medicina Legal, y Psiquiatría*. By 1903 the title had changed to *Archivos de Psiquiatría, Criminología, y Ciencias Afines*. (The new title was based on the Italian journal *Archivio di Psichiatria e Antropologia Criminale*, which had been founded in 1880 by Cesare Lombroso, Enrico Ferri, and Raffaele Garofalo.) In 1914 it was replaced by the *Revista de Criminología, Psiquiatría, y Medicina Legal*, edited by Helvio Fernández, and ran uninterrupted until 1935.

9. José Ingenieros, *Instituto de Criminología. Fundado en 1907* (Buenos Aires: Talleres Gráficos de la Penitenciaría Nacional, 1911), 14.

10. See Nancy Leys Stepan, "Race, Gender, and Nation in Argentina: The Influence of Italian Eugenics," *History of European Ideas*, 15, no. 4–6 (1992): 749–56.

11. José Ingenieros, *APCCA* 6 (1907): 5.

12. José Ingenieros, *Criminología* (1907; 6th ed., Buenos Aires: L. J. Rosso, 1916), 284.

13. José Ingenieros, "Relaciones de la psiquiatría con la psicología," *APCCA* 1 (1902): 500.

14. Francisco del Greco, "La psicopatología sintética y aplicada," *APCCA* 5 (1906): 1–2.

15. José Ingenieros, "Valor de la psicopatología en la antropología criminal," *APCCA* 1 (1902): 5.

16. José Ingenieros, "Las teorias de Lombroso ante la critica," *APCCA* 1 (1902): 337. See Cesare Lombroso, *L'uomo di genio* (1863; 5th ed., Torino: Bocca, 1888); and José María Ramos Mejía, *Las neurosis de los hombres célebres en la historia argentina* (Buenos Aires: M. Biedma, 1878).

17. José Ingenieros, "Nueva clasificación de los delincuentes," *APCCA* 5 (1906): 33. Emphasis in original.

18. Ingenieros, "Valor de la psicopatología," 11. Emphasis in original; the spelling of "criminalogía," used until 1903, alludes to the scientific focus on the individual criminal.

19. José Ingenieros, "Curso de psicología. Programa," reprinted in Hugo Vezzetti, *El nacimiento de la psicología en la Argentina* (Buenos Aires: Puntosur, 1987), 173–77.

20. Ingenieros, *Criminología*, 285–86.

21. Ingenieros, "Nueva clasificación de los delincuentes," 30–39.

22. On rural crime in Argentina, see Richard W. Slatta, "Rural Criminality and Social Conflict in Nineteenth-Century Buenos Aires Province," *Hispanic American Historical Review* 60, no. 3 (1980): 450–72; and Mark D. Szuchman, "Dis-

order and Social Control in Buenos Aires, 1810–1860," *Journal of Interdisciplinary History* 15, no. 1 (1984): 83–110.

23. Lancelotti's statistics are discussed in Julia Kirk Blackwelder and Lyman L. Johnson, "Changing Criminal Patterns in Buenos Aires, 1890–1914," *Journal of Latin American Studies* 14, no. 2 (1982): 371, 367. See also Miguel A. Lancelotti, "La criminalidad en Buenos Aires, 1887 a 1912," *RCPML* 1 (1914).

24. J. A. González Lanuza, "A propósito de Lombroso y del tipo criminal," *APCCA* 5 (1906): 273–74. Emphasis in original.

25. Ingenieros, "Valor de la psicopatología," 6–7.

26. Juan Vucetich, *Dactiloscopia comparada: El nuevo sistema argentino* (La Plata: Jacobo Preuser, 1904), 55.

27. Drago, *Los hombres de presa*, 92.

28. Ibid., 161.

29. Eusebio Gómez, *La mala vida en Buenos Aires* (Buenos Aires: Juan Roldán, 1908).

30. *APCCA* 6 (1907): 109.

31. Francisco de Veyga, "El espíritu y el alcance de la obra de Lombroso," *APCCA* 5 (1906): 266.

32. Ibid., 266, 267–68.

33. González Lanuza, "A propósito de Lombroso y del tipo criminal," 284.

34. Veyga, "El espíritu y el alcance de la obra de Lombroso," 263.

35. Ingenieros, "Valor de la psicopatología," 1.

36. Ingenieros, *Instituto de Criminología*, 13.

37. José María Ramos Mejía, "La enseñanza de la neuropatología en 1902," *APCCA* 2 (1903): 22.

38. Ingenieros, *Instituto de Criminología*, 6.

39. Aline Helg, "Race in Argentina and Cuba, 1880–1930," in *The Idea of Race in Latin America, 1870–1940*, ed. Richard Graham (Austin: University of Texas Press, 1990), 42; Eduardo A. Zimmermann, "Racial Ideas and Social Reform: Argentina, 1890–1916," *Hispanic American Historical Review* 72, no. 1 (1992): 23–46.

40. Victor Mercante, "Estudio sobre criminalidad infantil," *APCCA* 4 (1905): 578.

41. Lucas Ayarragaray, "La constitución étnica argentina y sus problemas," *APCCA* 12 (1912): 23, 24, 25.

42. Marie y Viollet, "Frecuencia de las enfermedades nerviosas," *APCCA* 9 (1910): 116. See also Jacobo Rietti, "La circuncisión de los Judíos en la República Argentina," *APCCA* 1 (1902): 339–68; Marie y Viollet, "Antisemitismo y locura," *APCCA* 9 (1910): 116–18; and J. M. Grandjean, "Las prohibiciones alimenticias de los Hebreos," *APCCA* 12 (1913): 252–53.

43. Terán, *José Ingenieros*, 57.

44. "Cerrando un ciclo," *APCCA* 12 (1913): 642. See also "Programa de la 'Revista de Criminología, Psiquiatría, y Medicina Legal,'" *RCPML* 1 (1914): 3–5.

45. "Cerrando un ciclo," 641.

46. "Programa de la 'Revista de Criminología, Psiquiatría, y Medicina Legal,' " 3.

47. Ibid., 4.

48. Ingenieros, *Criminología*, 10.

Chapter Four

1. This description of Drago's work is in William Belmont Parker, *Argentines of To-Day*, 2 vols. (New York: The Hispanic Society of America, 1920), 1:11. See Luis María Drago, *Los hombres de presa* (Buenos Aires: Lajouane, 1888).

2. José Ingenieros, "Prólogo," in Eusebio Gómez, *La mala vida en Buenos Aires* (Buenos Aires: Juan Roldán, 1908), 5, 13.

3. J. Edward Chamberlin and Sander L. Gilman, eds., *Degeneration: The Dark Side of Progress* (New York: Columbia University Press, 1985), ix; Dain Borges, " 'Puffy, Ugly, Slothful, and Inert': Degeneration in Brazilian Social Thought, 1880–1940," *Journal of Latin American Studies* 25, no. 2 (1993): 235–56.

4. Francisco de Veyga and Juan C. Córdoba, "Degeneración psíquica en los delincuentes profesionales," *APCCA* 1 (1902): 500–501.

5. Silvio Tatti, "La pulsación del pié en los criminales y en los locos," *APCCA* 1 (1902): 440.

6. Benjamin Solari, "Degeneración i crimen" (Medical thesis, UBA, 1891), 19–20.

7. Cayetano Sobre-Casas, "La histeria en la ginecología" (Medical thesis, UBA, 1895), 40–41.

8. Unknown author ("Ll."), "Las funciones sexuales, la locura y el delito en la mujer," *APCCA* 1 (1902): 59.

9. José Ingenieros, *Histeria y sugestión. Ensayos de psicología clínica* (1904; 5th ed., Buenos Aires: L. J. Rosso, 1919), 97–98.

10. In addition to Ingenieros's book on hysteria, see Joaquín J. Durquet, "Paraplegia histérica. Curación por sugestión," *APCCA* 4 (1905): 306–18; Joaquín J. Durquet, "Manía ambulatoria epiléptica y monoplegia histérica," *APCCA* 5 (1906): 341; and "La nueva histeria," *APCCA* 9 (1910): 620–26.

11. See Julia Rodriguez, "The Argentine Hysteric: A Turn-of-the-Century Psychiatric Type," in *Argentina on the Couch: Psychiatry, State, and Society, 1880 to the Present*, ed. Mariano Plotkin (Albuquerque: University of New Mexico Press, 2003), 25–48; and Gabriela Nouzeilles, "An Imaginary Plague in Turn-of-the-Century Buenos Aires: Hysteria, Discipline, and Languages of the Body," in *Disease in the History of Modern Latin America: From Malaria to AIDS*, ed. Diego Armus (Durham, N.C.: Duke University Press, 2003), 51–75.

12. Diógenes Decoud, "Estudio del hipnotismo," *ACMA* 11, no. 1 (1888): 4–27, quotations on 5.

13. Ricardo Schatz, "Contribution to the Study of Hysterical Paralysis" (Medical thesis, UBA, 1891), 34.

14. Francisco Netri, "El histerismo en la criminalidad," *APCCA* 1 (1902): 145.

15. Nicolás Vaschide, "La psicofisiología de impulso sexual," *APCCA* 5 (1906): 417–27.

16. José Ingenieros, "Patología de las funciones psicosexuales. Nueva clasificación genética," *APCCA* 9 (1910): 21–22, 23.

17. Bernardo Etchepare, "Desequilibrio mental, morfinomanía, e histeria," *APCCA* 11 (1912): 717–18, 722.

18. Fernando Raffo, "Locura puerperal" (Medical thesis, UBA, 1888), 17–18.

19. Carlos Fonso Gandolfo, "Psicosis puerperales" (Medical thesis, UBA, 1917), 40.

20. Eliseo Cantón and José Ingenieros, "Locura del embarazo," *APCCA* 2 (1903): 548–49.

21. Adolfo Batíz, *Buenos Aires, la ribera y los prostíbulos en 1880. Contribución a los estudios sociales, libro rojo* (Buenos Aires: Ediciones AGA-TAURA, 1960), 85–86, 120. It is not known when Batíz's work was originally published, but it is likely that he was writing in the early 1900s.

22. Gómez, *La mala vida en Buenos Aires*, 122, 128–29, 135.

23. H. Leale, "La criminalidad de los sexos," *APCCA* 11 (1912): 378–82. Originally appeared in the French journal *Archives d'Anthropologie Criminelle*, No. 98.

24. Ibid., 378–79. See also Horacio Areco, review of *Les femmes homicides* by Pauline Tarnowski, *APCCA* 7 (1908): 382.

25. Victor Mercante, "La mujer moderna," *APCCA* 8 (1909): 333–49, quotation on 342.

26. Antonio Dellepiane, *Las causas del delito* (Buenos Aires: Pablo Coni, 1892), 263.

27. See Lila M. Caimari, "Whose Prisoners Are These? Church, State, and *Patronatos* and the Rehabilitation of Female Criminals (Buenos Aires, 1890–1970)," *Americas* 54, no. 2 (1997): 195.

28. Cupertino Del Campo, "La kleptomania," *APCCA* 4 (1905): 84, 85–87; J. Rius y Matas, "Nicotinmanía," *APCCA* 5 (1906): 344–48; Jorge E. Coll, "Un caso de cleptomanía," *RCPML* 3 (1916): 350–51.

29. Carlos F. Roche, "El pseudo-hermafrodismo masculino y los androginoides," *APCCA* 3 (1904): 427. See also Manuel Podestá, "Un caso de hermafrodismo," *ACMA* 10, no. 2 (1887): 43–52.

30. Ingenieros, "Patología de las funciones psicosexuales," 25. See, for example, the case study "Una tríbade uxoricida," in ibid., 29.

31. Victor Mercante, "Fetiquismo y uranismo feminino," *APCCA* 4 (1905): 25.

32. For an in-depth discussion of the medical and hygienic discourse on homosexuality, see Jorge Salessi, *Médicos maleantes y maricas. Higiene, criminología y homosexualidad en la construcción de la nación Argentina (Buenos Aires: 1871–1914)* (Buenos Aires: Beatríz Viterbo Editora, 1995), part 3.

33. Benigno Lugones, "Prodromo á una descripcion de la pederastía pasiva," *ACMA* 3, no. 1 (1879): 4–16, quotation on 13.

34. Guillermo Olano, "La secrección mamaria en los invertidos sexuales," *APCCA* 1 (1902): 306.

35. Francisco de Veyga, "Inversion sexual congénita," *APCCA* 1 (1902): 46.

36. Jorge Salessi, "The Argentine Dissemination of Homosexuality, 1890–1914," in *Entiendes? Queer Readings, Hispanic Writings*, ed. Emilie L. Bergmann and Paul Julian Smith (Durham, N.C.: Duke University Press, 1995), 78.

37. Francisco de Veyga, "El sentido moral y la conducta de los invertidos," *APCCA* 3 (1904): 28. Cited in Salessi, "Argentine Dissemination of Homosexuality," 79; my translation of the original.

38. Gómez, *La mala vida en Buenos Aires*, 178.

39. Donna Guy, "Prostitution and Female Criminality in Buenos Aires, 1875–1937," in *The Problem of Order in Changing Societies: Essays on Crime and Policing in Argentina and Uruguay, 1750–1940*, ed. Lyman L. Johnson (Albuquerque: University of New Mexico Press, 1990), 99.

40. Belisario J. Montero, "La regeneración de los mendigos y vagabundos," *APCCA* 1 (1902): 648.

41. Drago, *Los hombres de presa*, 48.

42. Cornelio Moyano Gacitúa, *La delincuencia argentina* (Cordoba: F. Domenici, 1905), 249–50.

43. Francisco de Veyga and Juan C. Córdoba, "Degeneración psíquica en los delincuentes profesionales," *APCCA* 1 (1902): 501. On vagabonds, see also Alfredo Niceforo, "Lignes générales d'une anthropologie des classes pauvres," *APCCA* 5 (1906): 385–416; and P. Consiglio, "Los vagabundos," *APCCA* 10 (1911): 432–49.

44. Agustin J. Drago and Obdulio Hernandez, "Informe presentado al senor juez de instrucción Dr. Servando A. Gallegos," *SM*, 14 April 1898, 117–19, quotation on 118.

45. Mary Gibson, *Born to Crime: Cesare Lombroso and the Origins of Biological Criminality* (Westport, Conn.: Praeger, 2002), 67–73; Caimiri, "Whose Prisoners Are These?" 195.

46. Dellepiane, *Las causas del delito*, 284–85, 262.

47. Drago, *Los hombres de presa*, 135.

48. Dellepiane, *Las causas del delito*, 256–57, 264; Antonio Dellepiane, *El idioma del delito* (Buenos Aires: Arnoldo Moen, 1894), 8–13.

49. Dellepiane, *El idioma del delito*, 36, 44.

50. Drago, *Los hombres de presa*, chap. 8, esp. 73–75.

51. José Ingenieros, *Locura, simulación y criminalidad* (Buenos Aires: Talleres Gráficos de la Penitenciaría Nacional, 1910), 7–20.

52. Ibid., 26–27, 30.

53. Francisco de Veyga, "Degeneración, locura y simulación en los ladrones profesionales," *APCCA* 1 (1902): 705.

54. Ibid., 708, 711.

55. Francisco de Veyga, "La simulación del delito," *APCCA* 5 (1906): 165, 172, 180.

56. José María Ramos Mejía, *Las multitudes argentinas* (1899; Buenos Aires: Editorial de Belgrano, 1977), 199–201.

57. Ibid., 33–34.

58. Robert Nye, *Crime, Madness, and Politics in Modern France: The Medical Concept of National Decline* (Princeton, N.J.: Princeton University Press, 1984), 180; Robert Nye, *Origins of Crowd Psychology: Gustave Le Bon and the Crisis of Mass Democracy in the Third Republic* (London: Sage Publications, 1975).

59. Rafael de Robertis, "El alma de la muchedumbre," *APCCA* 4 (1905): 743.

60. Juan Agustín García, *Introducción general al estudio de las ciencias sociales argentinas* (Buenos Aires: Pedro Igon, 1899), 41. Emphasis in original.

61. Ramos Mejía, *Las multitudes argentinas*, 27, 30.

62. García, *Introducción general al estudio de las ciencias sociales*, 41.

63. Ronaldo Munck, "Cycles of Class Struggle and the Making of the Working Class in Argentina, 1890–1920," *Journal of Latin American Studies* 19, no. 1 (1987): 24.

64. Moyano Gacitúa, *La delincuencia argentina*, 11, 3.

65. José Gregorio Rossi, "La delincuencia profesional en Buenos Aires," *APCCA* 2 (1903): 170.

66. Moyano Gacitúa, *La delincuencia argentina*, 11; Rossi, "La delincuencia profesional en Buenos Aires," 170.

67. Abel A. Sonnenberg, "Degeneración y delincuencia," (Medical thesis, UBA, 1912), 109–17, 162.

68. Ingenieros, "Prólogo," 6, 5, 13.

69. Helvio Fernández, "Criminología argentina," *APCCA* 11 (1912): 628.

70. Miguel A. Lancelotti, "La criminalidad en Buenos Aires, 1887 a 1912," *RCPML* 1 (1914): 129–31; on immigration, 146–47.

71. Borges, " 'Puffy, Ugly, Slothful, and Inert,' " 241.

72. Ramos Mejía, *Las multitudes argentinas*, 212.

73. Ramón Pacheco, *Los regicidios en 1905. Consideraciones médicolegales* (Buenos Aires: Agustín Etchepareborda, 1905), 8–9, 5–6.

74. Eusebio Gómez, "La mala vida en Buenos Aires," *APCCA* 6 (1907): 431–42, quotations on 439.

75. Francisco de Veyga, *Anarquismo y anarquistas. Estudio de antropología criminal* (Buenos Aires, 1897), 449, 451, 438, 448.

76. Francisco de Veyga, "Delito político. El anarquista Planas Virella," *APCCA* 5 (1906): 516–17.

77. Ibid., 540, 529–30.

78. Bernaldo de Quirós, "Psicología del crimen anarquista," *APCCA* 7 (1913): 122–26, quotations on 123.

79. José G. Angulo, "La nueva ciencia eugénica y la esterilización de los degenerados," *APCCA* 11 (1912): 625.

Chapter Five

1. Rafael Fragueiro, *La niña argentina* (Buenos Aires: Cabaut y Compañía, 1902), 7–8, 43–44.

2. On the naturalistic novels of the period that deal with social pathological themes, see David William Foster, *The Argentine Generation of 1880: Ideology and Cultural Texts* (Columbia: University of Missouri Press, 1990); and Gabriela Nouzeilles, *Ficciones somáticas. Naturalismo, nacionalismo y políticas médicas del cuerpo (Argentina 1880–1910)* (Buenos Aires: Viterbo, 2000).

3. See Lila M. Caimari, *Perón y la Iglesia Católica: Religión, estado y sociedad en la Argentina (1943–1955)* (Buenos Aires: Ariel Historia, 1994), 27–56.

4. Elvira V. López, "La mujer y la enseñanza industrial," *Estudios* 1 (1901), cited and translated in Eduardo A. Zimmermann, "Racial Ideas and Social Reform: Argentina, 1890–1916," *Hispanic American Historical Review* 72, no. 1 (1992): 40.

5. Victor Mercante, "La mujer moderna," *APCCA* 8 (1909): 348–49.

6. David Rock, *Argentina, 1516–1987* (Berkeley: University of California Press, 1987), 176; Donna Guy, *Sex and Danger in Buenos Aires: Prostitution, Family, and Nation in Argentina* (Lincoln: University of Nebraska Press, 1990), 43.

7. Héctor Recalde, *Mujer, condiciones de vida, trabajo, y salud* (Buenos Aires: Centro Editor América Latina, 1988).

8. *Primer censo de la República Argentina* (Buenos Aires: Imprenta del Porvenir, 1872), xlv, xlvii; James R. Scobie, *Buenos Aires: Plaza to Suburb, 1870–1910* (New York: Oxford University Press, 1974), 140–41.

9. Lucio V. López and Amador L. Lucero, "Demencia inicial é incapacidad civil," *APCCA* 9 (1910): 95.

10. Lucas Ayarragaray, "Sobre impotencia sexual," *APCCA* 2 (1903): 270.

11. *La Voz de la Mujer*, 18 October 1896.

12. Marcela Nari, "Las practices anticonceptivas, la disminución de la natalidad y debate médico, 1890–1940," in *Política, médicos y enfermedades. Lecturas de historia de salud en la Argentina*, ed. Mirta Zaida Lobato (Mar del Plata: Editorial Biblos, 1996), 153–92.

13. Asunción Lavrin, *Women, Feminism, and Social Change in Argentina, Chile, and Uruguay, 1890–1940* (Lincoln: University of Nebraska Press, 1995), 128.

14. José Ingenieros, "Patología de las funciones psicosexuales. Nueva clasificación genética," *APCCA* 9 (1910): 35–36.

15. Carlos Baires, "La impotencia sexual como causa de divorcio," *APCCA* 8 (1909): 641–70, quotation on 656.

16. Ayarragaray, "Sobre impotencia sexual," 270.

17. Baires, "La impotencia sexual," 656–57. Emphasis in original.

18. Ingenieros, "Patología de las funciones psicosexuales," 36. Some examples of studies focusing on female sexual pathology include J. Roux, "La castración de la mujer y los deseos sexuales," *APCCA* 2 (1903): 50; R. Palacios, "Un caso de embarazo nervioso sugestivo," *APCCA* 2 (1903): 50–52; Eliseo Cantón and José

Ingenieros, "Locura del embarazo," *APCCA* 2 (1903): 548; and Pedro I. Oro, "Consideraciones sobre psicosis puerperal," *APCCA* 5 (1906): 582–605.

19. Ingenieros, *Histeria y sugestión*, 81–82, 71, 75, 98.

20. Ibid., 98.

21. Victor Arreguine, "El suicidio," *APCCA* 4 (1905): 706.

22. Ernesto Quesada, "Sobre cumplimiento de los deberes matrimoniales," *APCCA* 6 (1907): 221.

23. Antonio Dellepiane, *Las causas del delito* (Buenos Aires: Pablo Coni, 1892), 182, 198–99.

24. Barthou, "Sobre el divorcio," *APCCA* 2 (1903): 374.

25. Kristin Ruggiero, "Houses of Deposit and the Exclusion of Women in Turn-of-the-Century Argentina," in *Isolation: Places and Practices of Exclusion*, ed. Carolyn Strange and Alison Bashford (New York: Routledge, 2003), 119–32.

26. Maria Abella de Ramírez cited and translated in Lavrin, *Women, Feminism, and Social Change*, 34.

27. Cited in Recalde, *Mujer, condiciones de vida, trabajo, y salud*, 22–23.

28. *Primer censo de la República Argentina*, xlv, xlvii; Emilio Coni, *Higiene aplicada. La provincia de Corrientes* (Buenos Aires: Pablo Coni, 1898), 343–46; Guy, *Sex and Danger in Buenos Aires*, 68. Coni's *Higiene aplicada* was praised by a reviewer who noted that Coni's immediate goal was the establishment of a system of "sanitary police" (*SM*, 30 June 1898, 440).

29. Guy, *Sex and Danger in Buenos Aires*, chap. 1.

30. Enrique Prins, "Sobre la prostitución en Buenos Aires," *APCCA* 2 (1903): 725; Guy, *Sex and Danger in Buenos Aires*, 52–53.

31. Ibid., 721; Enrique Revilla, "El ejercicio de la prostitución en Buenos Aires," *APCCA* 2 (1903): 74; Francisco Sicardi, "La vida del delito y de la prostitución," *APCCA* 2 (1903): 11. On the criminological debates over prostitution, see Guy, *Sex and Danger in Buenos Aires*, 77–94.

32. See "El sexo," chap. 5 of Dellepiane, *Las causas del delito*.

33. Enrique Feinmann, *Policía social. Estudio sobre las costumbres y la moralidad pública* (Buenos Aires: Imprenta y Encuadernación de la Policía, 1913), 49.

34. Adolfo Batíz, *Buenos Aires, la ribera y los prostíbulos en 1880. Contribución a los estudios sociales, libro rojo* (Buenos Aires: Ediciones AGA-TAURA, 1960), 85–86, 120.

35. Feinmann, *Policía social*, 79.

36. Prins, "Sobre la prostitución en Buenos Aires," 722–26.

37. On control of prostitutes' movements, see Guy, *Sex and Danger in Buenos Aires*, 43, 61, 83–85.

38. Feinmann, *Policía social*, 31–32.

39. Kristin Ruggiero, "Honor, Maternity, and the Disciplining of Women: Infanticide in Late Nineteenth-Century Buenos Aires," *Hispanic American Historical Review* 72, no. 3 (1992): 354.

40. Ingenieros, "Patología de las funciones psicosexuales," 57.

41. Kristin Ruggiero, *Modernity in the Flesh: Medicine, Law, and Society in Turn-of-the-Century Argentina* (Stanford, Calif.: Stanford University Press, 2004), 56; Guy, *Sex and Danger in Buenos Aires*, 57–59.

42. Antonio Sagarna, "Absolución de una infanticide," *APCCA* 11 (1912): 167–72.

43. Ruggiero, *Modernity in the Flesh*, 56–61.

44. José Ingenieros, *Tratado del amor. La metafísica del amor, teoría genética del amor, psicología del amor* (4th ed.; Buenos Aires: Losada, 1990). See also Diego Carbonell, "El amor en biología," *RCPML* 2 (1915): 363–67.

45. Ingenieros, "Patología de las funciones psicosexuales," 14, 15, 16, 40.

46. Carlos Benitez and Juan Acuña, "Locura histerica," *APCCA* 2 (1903): 214.

47. Ibid., 219, 212.

48. *Tercer censo nacional de la República Argentina*, 10 vols. (Buenos Aires: Rosso y Compañía, 1916), 1:280–83.

49. For census figures on marriage rates, birthrates, and fertility, see ibid., 10:283–300.

50. Juan Bialet Massé, *Informe sobre el estado de la clase obrera* (1904; Buenos Aires: Hyspamérica, 1986), 650.

51. Ibid., 654–55, 1013; Cornelio Moyano Gacitúa, *La delicuencia argentina* (Cordoba: F. Domenici, 1905), 69.

52. L. Aubanel, "Hijos legítimos y hijos naturales," *APCCA* 2 (1903): 180.

53. Review of Hans W. Gruhle, "La infancia abandonada. Sus causas y sus raciones con la criminalidad," *APCCA* 12 (1913): 242–43.

54. H. A. Kozloff, *SM*, 16 February 1899, 64.

55. G. von Bunge, "De la impotencia creciente de las mujeres para amamantar a sus hijos," *RCPML* 2 (1916): 80–81.

56. Miguel A. Lancelotti, "La criminalidad en Buenos Aires, 1887 a 1912," *RCPML* 1 (1914): 28.

57. Eusebio Gómez, "Ladrones y usuarios," chap. 3 of *La mala vida en Buenos Aires* (Buenos Aires: Juan Roldán, 1908).

58. "La reincidencia. Medios para combatirla" (Letter from J. L. Duffy to Joaquín V. González), *RPen* 1, no. 1 (1905): 27.

59. "Estudios médico-legales. Informe del director del cuerpo médico en la Cárcel de Encausados," *RPen* 4, no. 1 (1908): 17–19.

60. Santín Carlos Rossi, "La salud y mentalidad de los niños en relación con las condiciones económicas del hogar," *RCPML* 7 (1920): 123–24.

61. H. Leale, "La criminalidad de los sexos," *APCCA* 11 (1912): 378–82.

62. Victor R. Pesenti, "Influencia de la civilización sobre el movimiento de la criminalidad" (Law thesis, UBA, 1901), 57–58.

63. Leale, "La criminalidad de los sexos," 381–82.

64. Pesenti, "Influencia de la civilización"; Aubanel, "Hijos legítimos e hijos naturales," 185.

65. Victoria Mazzeo, *Mortalidad infantil en la ciudad de Buenos Aires* (Buenos Aires: América Latina, 1993), 23.

66. Genaro Sisto, "Prólogo," in Enrique Feinmann, *La ciencia del niño* (Buenos Aires: Cabaut y Compañía, 1915), v.

67. Lavrin, *Women, Feminism, and Social Change*, 109–20.

68. Cited and translated in ibid., 87.

69. Donna J. Guy, "Emilio and Gabriela Coni: Reformers, Public Health, and Working Women," in *The Human Tradition in Latin America: The Nineteenth Century*, ed. William Beezley and Judith Ewell (Wilmington, Del.: SR Books, 1989), 233–47.

70. Enrique Feinmann, *Profilaxis social del delito* (Buenos Aires: Coni Hermanos, 1913), 20, 18.

71. Mazzeo, *Mortalidad infantil*, 70–71.

72. Feinmann, *La ciencia del niño*, 5.

Chapter Six

1. "Noticias de policía. El Descanso Dominical Obligatorio. Resolución del jefe de policía. Una campaña contra el alcoholismo," *La Prensa*, 11 February 1911.

2. Juan Bialet Massé, *Informe sobre el estado de la clase obrera* (1904; Buenos Aires: Hyspamérica, 1986), 999.

3. Donna Guy, *Sex and Danger in Buenos Aires: Prostitution, Family, and Nation in Argentina* (Lincoln: University of Nebraska Press, 1990), chap. 5.

4. *Anuario Estadístico de la Ciudad de Buenos Aires* 15 (1905): 232–33; *General Census of the Population, Buildings, Trades, and Industries of the City of Buenos Aires*, 3 vols. (Buenos Aires: Compañia Sud-Americana de Billetes de Banco, 1910), 2:303.

5. "Mendicidad y vagancia. Sus relaciones con la criminalidad," *RP*, 15 October 1899, 402–7, cited in Beatriz Ruibal, *Ideología del control social: Buenos Aires, 1880–1920* (Buenos Aires: Centro Editor de América Latina, 1993), 40.

6. Belisario J. Montero, "El parasitismo social y la beneficiencia pública," *APCCA* 3 (1904): 589.

7. Pietro Gori, "Alcoholismo y criminalidad en Chile," *APCCA* 1 (1902): 31–32.

8. Luís Martín Istúriz, "Influencia del alcohol sobre los trastornos mentales," *APCCA* 5 (1906): 253.

9. Fermín Rodríguez, "Influencia del alcoholismo sobre el suicidio, en Buenos Aires," *APCCA* 4 (1905): 547; Joaquín Durquet, "Observaciones sobre clínica psiquiátrica," *APCCA* 5 (1906): 736.

10. Durquet, "Observaciones sobre clínica psiquiátrica," 736.

11. G. von Bunge, "Las fuentes de la degeneración," *APCCA* 12 (1913): 480. Emphasis in original.

12. Francisco Netri, "La guerra al delito y la crísis del derecho penal," *APCCA* 5 (1906): 68.

13. Héctor A. Taborda, "Conferencias contra el alcoholismo," *APCCA* 6 (1907): 683.

14. Von Bunge, "Las fuentes de la degeneración," 483.

15. José Ingenieros and Juan C. Córdoba, "La defensa social y los alcoholistas crónicos," *APCCA* 2 (1903): 89.

16. Lucas Ayarragaray and José Ingenieros, "Demencia alcohólica y incapacidad civil," *APCCA* 2 (1903): 622.

17. Leonidas Avendaño, "Legislación antialcohólica en el Perú," *APCCA* 1 (1902): 490.

18. Joaquín Castro Soffia, "La locura masculina en Chile en 1901," *APCCA* 1 (1902): 726.

19. Review of F. García y Santos, *La degeneración de la raza por el alcohol* (Montevideo: Barreiro y Ramos, 1902), *APCCA* 1 (1902): 313.

20. Enrique Feinmann, *Policía social. Estudio sobre las costumbres y la moralidad pública* (Buenos Aires: Imprenta y Encuadernación de la Policía, 1913), 4.

21. Ibid., 13–14, 19–20, 24.

22. Von Bunge, "Las fuentes de la degeneración," 493.

23. Carlos Bunge, *Nuestra América. Ensayo de psicología social* (1903; Buenos Aires: Rosso y Compañía, 1918), 216, 264.

24. Montero, "El parasitismo social," 589.

25. P. Consiglio, "Los vagabundos," *APCCA* 10 (1911): 444, 447–48, 427–38.

26. *General Census of the Population, Buildings, Trades, and Industries*, 1: 396, 406. English in original.

27. Francisco de Veyga, *Los lunfardos* (Buenos Aires: Taller de la Penitenciaría Nacional, 1910), 17, 21.

28. Montero, "El parasitismo social," 586–87.

29. Crime statistics tabulated from the official statistical publication *Anuario Estadístico de la Ciudad de Buenos Aires* 1 (1891); 15 (1905); and 23 (1914). See also Alberto Méndez Casariego, *La criminalidad de la ciudad de Buenos Aires en 1887. Contravenciones, suicidios, y accidentes* (Buenos Aires: Imprenta del Departamento de Policía de la Capital, 1888); and the *Boletín mensual de estadística* (1895–99). These police publications included factors of nationality, age, marital status, and literacy.

30. José María Ramos Mejía, *Las multitudes argentinas* (1899; Buenos Aires: Editorial de Belgrano, 1977), 17–18, 27–28.

31. Ibid., 211–12.

32. *Copiador de Notas*, 1873–76, section 7, no. 2, 19 May 1876–21 October 1876, p. 266; *Copiador de Notas*, 1909, section 25, no. 84, 10 May 1909–9 June 1909, p. 425; *Copiador de Notas*, 1905, section 28, no. 72, 11 January 1905–13 March 1905, p. 363; all in Archivo de la Policía de la Capital, División Investigaciones, Buenos Aires.

33. *Memoria del departamento de Policía de la Capital, 1883–1884* (Buenos Aires: La Tribuna Nacional, 1884), 81.

34. "Expulsión de extranjeros," *RP*, 1 July 1899, 35–38, 53–56.

35. Guillermo J. Nunes, *Memoria del departamento de policía correspondiente al año 1891* (La Plata: Talleres del Museo de La Plata, 1892), 5.

36. James Scobie, *Argentina: A City and a Nation* (New York: Oxford University Press, 1964), 156–58.

37. *La Nación*, 23 January 1912.

38. *Caras y Caretas*, 12 November 1904.

39. *Memoria del departamento de la Policía de la Capital, 1906–1909* (Buenos Aires: Imprenta y Encuadernación de la Policía, 1909), 15–16.

40. Luis M. Doyenhard, *La policía en Sud-América* (La Plata: La Popular, 1905), part 1, 5–29.

41. Antonio Ballvé, *Texto de instrucción policial* (Buenos Aires: Talleres Gráficos de la Penitenciaría Nacional, 1907), cited in Antonio Ballvé, "Reglas generales de procedimiento policial en los delitos públicos," *APCCA* 6 (1907): 671–72.

42. *Memoria del Ministerio del Interior, 1880* (Buenos Aires: Imprenta Especial de Obras, 1881), 65–67; *Memoria del Ministerio del Interior, 1901–1904* (Buenos Aires, 1904), 12.

43. *Memoria del Ministerio del Interior, 1901–1904*, 11–13; *Memoria del Ministerio del Interior, 1912* (Buenos Aires, 1913), 231.

44. A. Cutrera, review of *La policía científica*, by Alfredo Niceforo, *APCCA* 6 (1907): 499.

45. *La Prensa*, 9 December 1911.

46. *La Prensa*, 1 June 1912.

47. *La Vanguardia*, 19 February 1909.

48. *La Voz de la Mujer*, 31 January 1896, 3.

49. *La Vanguardia*, 4 March 1905, 27 May 1905, 8 September 1905, 26 September 1905, 6 December 1905, 9 January 1907.

50. *Memoria del departamento de la Policía de la Capital, 1906–1909*, 15–16.

51. *Memoria del Ministerio del Interior, 1895* (Buenos Aires: Imprenta "la Tribuna," 1896), 560–61.

52. *Memoria del Ministerio del Interior, 1900* (Buenos Aires: Imprenta "la Tribuna," 1901), 114.

53. *Memoria del Ministerio del Interior, 1901–1904*, 103–6.

54. Ramón Falcón to Marco Avellaneda, 8 November 1907, bundle 11, file 2054 (1908); Ramón Falcón to Marco Avellaneda, 31 March 1908, bundle 8, file 1596-P (1908); both in series "Ministerio del Interior," AGN.

55. *Memoria del Ministerio del Interior, 1910* (Buenos Aires, 1911), 198–201, 217–22.

56. *Memoria del Ministerio del Interior, 1912*, 221–25.

57. *Memoria del Ministerio del Interior, 1887* (Buenos Aires: Imprenta "Sudamerica," 1888), 341.

58. See Julia Rodriguez, "South Atlantic Crossings: Fingerprints, Science, and the State in Turn-of-the-Century Argentina," *American Historical Review* 109, no. 2 (2004): 387–416.

59. Juan Vucetich to Luis M. Doyenhard, 1 May 1903, correspondence folder, AV.

60. Juan Vucetich, *Dactiloscopía comparada. El nuevo sistema argentino* (La Plata: Jacobo Preuser, 1904). An earlier publication, the pamphlet entitled *Instruc-*

ciones generales para la identificación antropométrica was published in 1893. Large portions of that pamphlet were reprinted in the La Plata newspaper *El Día*, 22 December 1893.

61. Vucetich, *Dactiloscopía comparada*, 41.
62. Juan Vucetich to Luis M. Doyenhard, 1 May 1903, correspondence folder, AV.
63. *Memoria del Ministerio del Interior, 1893* (Buenos Aires: Imprenta "la Tribuna," 1894), 257; *Memoria del Ministerio del Interior, 1899* (Buenos Aires: Tip. de la Penitenciaría Nacional, 1900), 52.
64. Bibiano S. Torres, *Observaciones sobre las Oficinas de Antropometría é Icnofalangometría* (Buenos Aires, 1894), 5, 9.
65. J. G. Rossi to Francisco Beazley, 9 September 1901, miscellaneous correspondence folder, AV. See also *Orden del Día*, 9 November 1903, 1066.
66. Belisario Otamendi to Juan Vucetich, 11 May 1901, miscellaneous correspondence folder, AV.
67. Juan Vucetich, "Historia sintética de la identificación," *Revista de Identificación y Ciencas Penales* 7 (1931): 69.
68. Ballvé, "Reglas generales de procedimiento policial en los delitos públicos," 678; José G. Rossi to Juan Vucetich, telegram, 31 December 1907, miscellaneous correspondence folder, AV.
69. *Memoria del departamento de la Policía de la Capital, 1906–1909*, 298. In 1911 a legislative committee called Vucetich to share his method at a special meeting with the minister of the interior and the minister of war. José Tonrouge to Juan Vucetich, 29 April 1911, correspondence folder "A–M," AV.
70. *Memoria de la división técnica, año 1916* (Buenos Aires: Imprenta y Encuadernación de la Policía, 1917), 18. City population figure from the 1916 census; see *Tercer censo nacional de la República Argentina*, 10 vols. (Buenos Aires: Rosso y Compañía, 1916), 1:3.
71. For detailed accounts of the street confrontations of 1909 and 1910, see Osvaldo Bayer, "Simón Radowisky," in *The Argentina Reader: History, Culture, Politics*, ed. Gabriela Nouzeilles and Graciela Montaldo (Durham, N.C.: Duke University Press, 2002), 222.
72. *Memoria del Ministerio del Interior, 1910*, 19–20.

Chapter Seven

1. Helvio Fernández, "Degeneración hereditaria con perturbaciones del sentido moral," *APCCA* 11 (1912): 758, 760.
2. On mid-nineteenth-century disciplinary regimes, see Ricardo Salvatore, *Wandering Paysanos: State Order and Subaltern Experience in Buenos Aires during the Rosas Period* (Durham, N.C.: Duke University Press, 2003), chap. 7.
3. Fermín Alsina, "Sistema penitenciario" (Law thesis, UBA, 1877), 10, 18, 23.
4. Editorial note to Joaquín V. González, "Tendencias modernas del sistema

penitenciario," *APCCA* 3 (1904): 641; Eusebio Gómez, *Estudios penitenciarios* (Buenos Aires: Talleres Gráficos de la Penitenciaría Nacional, 1906), 55.

5. Tomás R. Cullen, "Discurso del ministro de justicia," *RCPML* 1 (1914): 267.

6. Rodolfo Rivarola, "Discurso del delegado del gobierno nacional," *RCPML* 1 (1914): 263–64.

7. "El delito. Sus causas y sus remedies," *RPen* 2 (1906): 7.

8. "Estudios médico-legales. Informe del director del cuerpo médico en la Cárcel de Encausados," *RPen* 4, no. 1 (1908): 8, 24–27.

9. Rafael Súnico, "Reglamentación de la pena," *RCPML* 1 (1914): 352–53.

10. Rivarola, "Discurso del delegado del gobierno nacional," 266; Eusebio Gómez, "La función social de la pena," *RCPMP* 1 (1914): 362.

11. Gómez, "La función social de la pena," 355.

12. "Estudios médico-legales. Informe del director del cuerpo médico en la Cárcel de Encausados," 8.

13. Jonathan Ablard, "Law, Medicine, and Confinement to Public Psychiatric Hospitals in Twentieth-Century Argentina," in *Argentina on the Couch: Psychiatry, State, and Society, 1880 to the Present*, ed. Mariano Plotkin (Albuquerque: University of New Mexico Press, 2003), 87–112.

14. Ibid., 100–109.

15. Korn cited in Hugo Vezzetti, *La locura en Argentina* (Buenos Aires: Paidos, 1985), 143.

16. José Ingenieros, *Simulación de locura* (Buenos Aires: Elmer Editor, 1956), 91; Ablard, "Law, Medicine, and Confinement," 100.

17. José Ingenieros, *La locura en Argentina* (Buenos Aires, 1920), 234–35.

18. "La locura en las prisiones," *RPen* 2 (1906): 113–18.

19. Lila M. Caimari, "Whose Prisoners Are These? Church, State, and *Patronatos* and the Rehabilitation of Female Criminals (Buenos Aires, 1890–1970)," *Americas* 54, no. 2 (1997): 198, 192. On girls in the Women's Correctional House, see Donna Guy, "Girls in Prison: The Role of the Buenos Aires Casa Correccional de Mujeres as an Institution for Child Rescue, 1890–1940," in *Crime and Punishment in Latin America: Law and Society Since Late Colonial Times*, ed. Ricardo Salvatore, Carlos Aguirre, and Gilbert Joseph (Durham, N.C.: Duke University Press, 2001), 369–90.

20. Caimari, "Whose Prisoners Are These?" 195.

21. Eusebio Gómez, *Criminología argentina* (Buenos Aires: Imprenta Europea, 1912), x.

22. Kristin Ruggiero, "Wives on 'Deposit': Internment and the Preservation of Husbands' Honor in Late Nineteenth-Century Buenos Aires," *Journal of Family History* 17, no. 3 (1992): 253–70.

23. *Memoria del Ministerio de Justicia (1880)* (Buenos Aires: Talleres Gráficos de la Penitenciaría Nacional, 1881), 669.

24. *General Census of the Population, Buildings, Trades, and Industries of the City of*

Buenos Aires, 3 vols.(Buenos Aires: Compañia Sud-Americana de Billetes de Banco, 1910), 1: 394.

25. Gómez, *Estudios penitenciarios*, 15–16.

26. "La Cárcel de Encausados. Sus nuevos rumbos" (Letter from José Luis Duffy to Dr. Joaquín V. González, minister of justice), *RPen* 1, no. 1 (1905): 3.

27. Gómez, *Estudios penitenciarios*, 90, 87. Emphasis in original.

28. "El estudio del delincuente," *RPen* 4 (1908): 96–97.

29. Gómez, *Estudios penitenciarios*, 79–82, 69, 72.

30. "Oficina médica y enfermería. Reglamento interno," *RPen* 1 (1905): 200–202.

31. "Cárcel de Encausados. Estudios médico-legales. Cuaderno medico-psicológico," *RPen* 3, no. 1 (1907): 74–75.

32. "Oficina de estudios médico-legales," *RPen* 1 (1905): 36.

33. Gómez, *Estudios penitenciarios*, 81, 69–70.

34. "Oficina de estudios médico-legales," 32–34.

35. "Estudios médico-legales. Informe del director del cuerpo médico en la Cárcel de Encausados," *RPen* 4, no. 1 (1908): 7–8, 11, 15, 16.

36. See Ricardo D. Salvatore and Carlos Aguirre, eds., *The Birth of the Penitentiary in Latin America: Essays on Criminology, Prison Reform, and Social Control, 1830–1940* (Austin: University of Texas Press, 1996), 11.

37. Decree [*decreto*], 11 July 1869, reprinted in Eusebio Gómez, *Memoria descriptiva de la Penitenciaría Nacional de Buenos Aires* (Buenos Aires: Talleres Gráficos de la Penitenciaría Nacional, 1914), 36–43, quotations on 39, 40.

38. *Memoria del Ministerio de Justicia (1880)*, 657–58.

39. Antonio Ballvé, *La Penitenciaría Nacional de Buenos Aires. Conferencia leída en el Ateneo de Montevideo, el 22 de marzo de 1907, bajo el patrocinio del tercer congreso médico Latino Americano* (Buenos Aires: Talleres Gráficos de la Penitenciaría Nacional, 1907), 13, 18–19.

40. Ibid., 18–19, 21.

41. "Las cárceles modernas," *Caras y Caretas*, 10 August 1907.

42. *Memoria del Ministerio de Justicia (1881)* (Buenos Aires: Imprenta La Penitenciaría, 1882), xxxiii.

43. Gómez, *Estudios penitenciarios*, 56–57, 60–61.

44. "La nueva revista. Su programa." *RPen* 1, no. 1 (1905): 2–3.

45. "Las cárceles en la provincia," *La Vanguardia*, 19 September 1905.

46. Gómez, "La función social de la pena," 362.

47. Eusebio Gómez, "Cárceles y establecimientos," 297.

48. Francisco de Veyga, *Los lunfardos* (Buenos Aires: Taller de la Penitenciaría Nacional, 1910), 29.

49. *La Vanguardia*, 25 February 1905. The number of prisoners was calculated from the *Anuario Estadístico de la Ciudad de Buenos Aires* 13 (1903).

50. *La Vanguardia*, 8 September 1902.

51. "La nueva revista," 1–2. González's proposed legislation was cited in Gómez, *Estudios penitenciarios*, 98.

52. José Ingenieros, *Instituto de Criminología. Fundado en 1907* (Buenos Aires: Talleres Gráficos de la Penitenciaría Nacional, 1911), 258.

53. Ibid., 258–59.

54. Gina Lombroso, *L'Avanti* (Rome), 20 November 1907, reprinted in *APCCA* 7 (1908): 232, 236; Enrico Ferri, "La scuola criminale positiva," *Corriere della Sera* (Milan), reprinted in *APCCA* 7 (1908): 237.

55. Ingenieros, *Instituto de Criminología*, 257.

56. Horacio Areco, "Enrique Ferri y el positivismo penal," *APCCA* 7 (1908): 437.

57. Antonio Ballvé to Federico Pinedo, minister of justice, 4 May 1907, in *APCCA* 6 (1907): 260–61.

58. Gómez, *La Penitenciaría Nacional*, 78–79.

59. Ibid., 76, 93–94.

60. Antonio Ballvé, "Reglamentando la instalación y funciones del Instituto de Criminología," *APCCA* 6 (1907): 262.

61. Gómez, *Estudios penitenciarios*, 15–21, 28–29, 25–26, 94.

62. José Ingenieros, *Locura, simulación y criminalidad* (Buenos Aires: Talleres Gráficos de la Penitenciaría Nacional, 1910), 28.

63. Ibid., 31.

64. Gómez, *Estudios penitenciarios*, 12–14.

65. Diego González, A. Claros, and C. Muratgia, *Proyecto de reformas carcelarias. Informe de la comisión especial* (Buenos Aires: Talleres Gráficos de la Penitenciaría Nacional, 1913), 10.

66. Ballvé, *La Penitenciaría Nacional de Buenos Aires*, 109–10.

67. Gómez, *Estudios penitenciarios*, 55.

68. Cullen, "Discurso del ministro de justicia," 269.

69. Ballvé, *La Penitenciaría Nacional de Buenos Aires*, 123.

70. Gómez, "La función social de la pena," 357.

71. Ibid., 355.

72. Ibid., 355.

73. Gómez, "La función social de la pena," 358.

74. Gómez, *Estudios penitenciarios*, 56.

75. Gómez, *La Penitenciaría Nacional*, 63–64.

76. *Anuario Estadístico de la Ciudad de Buenos Aires* 1 (1891): 498.

77. Ballvé, *La Penitenciaría Nacional de Buenos Aires*, 112–20.

78. Gómez, *La Penitenciaría Nacional*, 51.

79. Ballvé, *La Penitenciaría Nacional de Buenos Aires*, 109–10.

80. Gómez, "La función social de la pena," 358–59.

81. Gómez, *Estudios penitenciarios*, 36, 34–41.

82. Gómez, "La funcion social de la pena," 340.

83. Gómez, *La Penitenciaría Nacional*, 63.

84. Antonio Ballvé to Federico Pinedo, minister of justice, 4 May 1907, in *APCCA* 6 (1907): 260–61.

85. Cornelio Moyano Gacitúa, *La delincuencia argentina* (Cordoba: F. Domenici, 1905), 410.

86. "La nueva revista," 2.

87. Gómez, "La función social de la pena," 358.

88. Cullen, "Discurso del ministro de justicia," 269.

89. Norberto Piñero, "Discurso," *RCPML* 1 (1914): 260.

90. Ballvé, *La Penitenciaría Nacional de Buenos Aires*, 87, 132.

91. Gómez, *Estudios penitenciarios*, 46.

92. "La nueva revista," 2.

Chapter Eight

1. For detail on Buenos Aires's expansion in this period, see James R. Scobie, *Buenos Aires: Plaza to Suburb, 1870–1910* (New York: Oxford University Press, 1974).

2. "La salud del pueblo es la suprema ley del Estado." Cited in Héctor Recalde, *Las epidemias de cólera (1856–1895). Salud y sociedad en la Argentina oligárquica* (Buenos Aires: Corregidor, 1993), 21, 13.

3. Jorge Salessi, *Médicos maleantes y maricas. Higiene, criminología, y homosexualidad en la construcción de la nación Argentina (Buenos Aires: 1871–1914)* (Buenos Aires: Beatríz Viterbo Editora, 1995), 97, 102.

4. José G. Angulo, "La nueva ciencia eugénica y la esterilización de los degenerados," *APCCA* 11 (1912): 625.

5. See Nancy Tomes, *The Gospel of Germs: Men, Women, and the Microbe in American Life* (Cambridge, Mass.: Harvard University Press, 1998).

6. Luis Agote, "Defensa sanitaria maritima contra las enfermedades exóticas viajeras," *ADNH* 8 (15 May 1898): 307, 308, 311.

7. Recalde, *Las epidemias de cólera*, 22.

8. Penna quoted in Héctor Recalde, *La salud de los trabajadores en Buenos Aires (1870–1910)* (Buenos Aires: Grupo Editor Universitaria, 1997), 294.

9. Scobie, *Buenos Aires*, 58.

10. Héctor Recalde, *Vida popular y salud en Buenos Aires (1900–1930)*, 2 vols. (Buenos Aires: Biblioteca Politica Argentina, 1994), 1:107–15; Recalde, *La salud de los trabajadores*, 203–46.

11. Donna Guy, *Sex and Danger in Buenos Aires: Prostitution, Family, and Nation in Argentina* (Lincoln: University of Nebraska Press, 1990), chap. 3.

12. Guillermo Rawson, *Conferencias sobre higiene pública dadas en la facultad de medicina de Buenos-Aires* (Paris: Donnamette and Hattu, 1876), 247, 230, 233.

13. Ricardo Gonzales Leandri, "La profesion médica en Buenos Aires: 1852–1870," in *Política, médicos y enfermedades. Lecturas de historia de la salud en la Argentina*, ed. Mirta Zaida Lobato (Buenos Aires: Editorial Biblos, 1996), 21–53. See also "Documentos oficiales," *RMQ* 4, no. 15 (8 November 1867): 226–27; and "La Junta de Higiene Pública," *RMQ* 5, no. 16 (23 November 1867): 242–43.

14. Decreed by Law No. 3379, Ministerio de Justicia, *Digesto de Justicia* 1 (1941): 315. See also Juan R. Fernandez, "Instituto de Medicina Legal y Morgue. En el proyecto de construcción de la Escuela Practica de Medicina," *APCCA* 1 (1902): 297.

15. Rawson, *Conferencias sobre hygiene pública*, 106.

16. Recalde, *Las epidemias de cólera*, 81.

17. Salessi, *Médicos maleantes y maricas*, 86–89.

18. Rawson, *Conferencias sobre higiene pública*, 257.

19. José Moya, *Cousins and Strangers: Spanish Immigrants in Buenos Aires, 1850–1930* (Berkeley: University of California Press, 1998), 172.

20. Scobie, *Buenos Aires*, 115, 149–50, 152, 154.

21. Quoted in Salessi, *Médicos maleantes y maricas*, 98–99; see also "Comisiones auxiliares de higiene," *ADNH* 2 (1892): 488.

22. Emilio R. Coni, *Progrès de l'hygiène dans la République Argentine* (Paris: Librairie J-B. Bailliere et Fils, 1887), 247–51, 262–64, 249–50.

23. Emilio R. Coni, *Código de higiene y medicina legal de la República Argentina para uso de médicos, abogados, farmacéuticos, etc.* (Buenos Aires: Librería de Juan Etchepareborda, 1891).

24. Law cited in Salessi, *Médicos maleantes y maricas*, 102.

25. "Higiene pública. La obra del Dr. Ramos Mejía en el Departamento Nacional de Higiene," *SM*, 20 October 1898, 349–50. See also "Departamento Nacional de Higiene," *SM*, 20 October 1898, dclxxv–dclxxvi.

26. Augusto Bunge, *Las conquistas de la higiene social. Informe presentado al excmo. gobierno nacional* (Buenos Aires: Penitenciaría Nacional, 1910), part 1.

27. Penna cited in Recalde, *La salud de los trabajadores*, 295, 296.

28. Cited in Recalde, *Las epidemias de cólera*, 92.

29. Ibid., 12, 24.

30. Carlos Pellegrini, "Inauguración del puerto de Buenos Aires," in *Pellegrini Obras*, ed. Agustín Rivero Astengo, 5 vols. (Buenos Aires: Coni, 1941), 3:52.

31. The following year, however, due to economic conditions, immigration dropped to 138,407. Juan A. Alsina, *La inmigración Europea en la República Argentina* (Buenos Aires, 1898), 125.

32. Agote, "Defensa sanitaria maritima," 307.

33. Amy L. Fairchild, *Science at the Borders: Immigrant Medical Inspection and the Shaping of the Modern Industrial Labor Force* (Baltimore, Md.: Johns Hopkins University Press, 2003).

34. Report to interior minister by commissioner general of immigration, 1 May 1876, in *La inmigración Europea en la Argentina*, ed. Alberto Kleiner, 10 vols. (Buenos Aires: Poligono, 1983), 1:20.

35. Ibid.

36. Miguel Cané, *Expulsión de extranjeros (proyecto de ley)* (Buenos Aires: J. Sarrailh, 1899), 125.

37. *DSCS*, 27 June 1910, 1:322.

38. Santiago Vaca-Guzmán, *La naturalización de los extranjeros* (Buenos Aires: Casa Editoria Jacobo Peuser, 1891), 60, 85–86.

39. "La inmigración como medio de progreso y de cultura," *Caras y Caretas*, 1 January 1914.

40. "Movimiento de la Oficina Nacional de Trabajo, 1875," in *La inmigración Europea en la Argentina*, ed. Kleiner, 4:59.

41. Agote, "Defensa sanitaria marítima," 307, 308, 311.

42. "Ley de Inmigración y Reglamento de Desembarco (Ley no. 817)," reprinted in Cayetano Carbonell, *Orden y trabajo*, 2 vols. (Buenos Aires: Lajouane, 1910), 2:4, 13.

43. Juan A. Alsina, "Breves consideraciones sobre la higiene del inmigrante" (Medical thesis, UBA, 1899), 15.

44. Ibid., 40, 44–45.

45. Agote cited in Olga Bordi de Ragucci, *Cólera y inmigración, 1880–1900* (Buenos Aires: Editorial Leviatán, 1992), 168–69.

46. Alsina, *La inmigración Europea en la República Argentina*, 125.

47. Alberto Kleiner, ed., *Publicidad oficial sobre Argentina como país de inmigración, 1903* (Buenos Aires: Poligono SRL, 1983).

48. *Memoria del Departamento General de Inmigración corespondiente al año 1890, presentada al Ministerio de Relaciones Exteriores* (Buenos Aires: M. Biedma, 1890), 75, 79.

49. Alsina wrote in 1898 that "it is not possible to take into account those who enter the country through our land borders, although the number is presently growing." Alsina, *La inmigración Europea en la República Argentina*, 124.

50. "Proyecto de modificaciones a la Ley de *Inmigración*," 22 June 1909, reprinted in Carbonell, *Orden y trabajo*, 2:28.

51. Alsina, "Breves consideraciones," 17.

52. Ibid., 18.

53. Report to interior minister by general commissioner of immigration, 1 May 1876, 1:12, 14.

54. Alsina, *La inmigración Europea en la República Argentina*, 122.

55. *Memoria del Departamento General de Inmigración . . . 1890*, 79.

56. Ibid., 79, 80.

57. Ibid.

58. *DSCS*, 27 June 1910, 1:325.

59. *DSCS*, 22 November 1902, 1:356, 297.

60. "Proyecto de modificaciones a la Ley de Inmigración," 22 June 1909, reprinted in Carbonell, *Orden y trabajo*, 2:27.

61. Carbonell, *Orden y trabajo*, 2:2, 3.

62. "Socialismo Argentino y legislación obrera," *APCCA* 11 (1912): 502–3.

63. Lucas Ayarragaray, "La constitución étnica argentina y sus problemas," *APCCA* 12 (1912): 23, 24, 25.

64. Carlos F. Gómez, *Ciudadanía y naturalización. Proyecto de ley* (Buenos Aires: Alsina, 1913), 29, 123–24.

65. José Ingenieros, *Criminología* (1907; 6th ed., Buenos Aires: L. J. Rosso, 1916), 287.

66. Luis Reyna Almandos, *Dactiloscopia argentina. Su historia e influencia en la legislación* (1909; La Plata:Universidad de la Plata, 1932), 125, 127.

67. Articles 226 to 250, *Proyecto de código penal para la República Argentina . . . (1891)* (2d ed., Buenos Aires: Penitenciaría Nacional, 1898), 276. For discussion, see pages 200–215, 276.

68. "Síntesis cronólogica," *APCCA* 11 (1912): xxxiv.

69. Dirección General de Immigración, "Resolución No. 292," 21 May 1912, general file (unclassified), AV.

70. Dirección General de Inmigración (Buenos Aires), "Libreta del inmigrado," 1911, 3, miscellaneous folder, AV.

71. *Proyecto de código penal (1891),* 213.

72. *Proyecto de código de procedimientos . . . (1913)* (Buenos Aires: Talleres Gráficos de la Penitenciaría Nacional, 1913), xlviii.

73. Francisco Durá, *Naturalización y expulsión de extranjeros* (Buenos Aires: Coni Hermanos, 1911), 280.

74. *La Prensa* published Cané's proposal in its entirety. See *La Prensa,* 9 June 1899, 4–5, 7.

Chapter Nine

1. Agustín Alvarez, *Manual de patología política* (Buenos Aires: Peuser, 1899), 4, 147.

2. Francisco Durá, *Naturalización y expulsión de extranjeros* (Buenos Aires: Coni Hermanos, 1911), 285.

3. The criminal code is uniform for the entire country; criminal procedure, such as organizational court matters and trial procedure, varies from province to province.

4. Rosa del Olmo, *América Latina y su criminología* (Mexico City: Siglo Veintiuno, 1987), 53.

5. Juan Ramón Fernández (minister of education), "Instituto de Medicina Legal y Morgue," *APCCA* 1 (1902): 296; Joaquín Gonzáles (interior minister), "La justicia Argentina y la muerte de Tallarico," *APCCA* 1 (1902): 654; Amador L. Lucero (national deputy), "Sobre un caso de locura sistematizada progresiva," *APCCA* 4 (1905): 45; Amador L. Lucero, "Desarrollo psícico y discernimiento para delinquir," *APCCA* 4 (1905): 464; Lucio V. López and Amador L. Lucero, "Demencia inicial é incapacidad civil," *APCCA* 9 (1910): 92; Jerónimo del Barco and I. Ruiz Moreno (both national deputies), "Proyecto sobre establecimientos penales," *APCCA* 7 (1908): 562.

6. Leonidas Avendaño, "Ejecución de penas en caso de enfermedad sobreviniente," *APCCA* 3 (1904): 449.

7. José Ingenieros, review of *La psychologie criminelle*, by P. Kowalewsky, *APCCA* 2 (1903): 121; Francisco de Veyga, "Sobre la prueba pericial. Contestacíon a la refutacíon del doctor Alba Carreras," *SM*, 2 June 1898, 190–91.

8. See Beatriz Ruibal, "Medicina legal y derecho penal a fines del siglo XIX," in *Política, médicos y enfermedades. Lecturas de historia de la salud en la Argentina*, ed. Mirta Zaida Lobato (Mar del Plata: Editorial Biblos, 1996), 193–207.

9. José María Ramos Mejía, "Informe médico-legal sobre el estado mental del procesado G.E.," *SM*, 11 August 1898, 266.

10. Samuel Gache, "Patogenía del suicidio en Buenos Aires," *ACMA* 7 (1883–84): 555–58.

11. The *APCCA* and *RCPML* published literally dozens of commentaries on the need for scientific reform of criminal law. Some examples include José L. Murature, "Cuestions penales de actualidad," *APCCA* 1 (1902): 12–17; Rodolfo Moreno, "Fundamentos de un proyecto de código penal," *RCPML* 3 (1916): 440–62; and "Proyecto del código penal, con las modificaciones introducidas por la Comisión de Códigos del Senado Nacional," *RCPML* 7 (1920): 186–236.

12. *Código penal de la República Argentina* (Buenos Aires: SudAmérica, 1887), 2.

13. *Proyecto de código penal para la República Argentina . . . (1891)* (2d ed., Buenos Aires: Penitenciaría Nacional, 1898), 48.

14. Ibid., 5, 47.

15. Carlos Saavedra et al., *Proyecto de código* (Buenos Aires: Tip. de la Cárcel de Encausados, 1906), 10–12.

16. Julio Herrera, *La reforma penal* (Buenos Aires: J. E. Hall, 1911), 342.

17. "Exposición de motivos," in Diego González, A. Claros, and C. Muratgia, *Proyecto de reformas carcelarias. Informe de la comisión especial* (Buenos Aires: Talleres Gráficos de la Penitenciaría Nacional, 1913), 162–63.

18. José Peco, *La reforma penal Argentina de 1917–20: Ante la ciencia penal contemporánea y los antecedentes nacionales y extranjeros* (Buenos Aires: V. Abelado, 1921), 386.

19. Argentina, Senado de la Nación, *La reforma penal en el Senado* (Buenos Aires: L. J. Rosso, 1919), 51–52.

20. Virgilio Ducceschi, "La criminología moderna," *APCCA* 12 (1913): 419–20; *Proyecto de código de procedimientos . . . (1913)* (Buenos Aires: Talleres Gráficos de la Penitenciaría Nacional, 1913), lix.

21. *Proyecto de código de procedimientos . . . (1913)*, xx, xlviii; *Proyecto de código penal para la República Argentina, . . . 1904* (Buenos Aires: Tipografía de la Cárcel de Encausados, 1906), lxvi.

22. Juan P. Ramos, letter to Rodolfo Moreno, in *Proyecto de código penal para la nación Argentina* (Buenos Aires: L. J. Rosso, 1917), 262.

23. Lila M. Caimari, "Whose Prisoners Are These? Church, State, and *Patronatos*

and the Rehabilitation of Female Criminals (Buenos Aires, 1890–1970)," *Americas* 54, no. 2 (1997): 200–201.

24. Dora Barrancos, "Inferioridad jurídica y encierro doméstico," in *Historia de las mujeres en la Argentina*, ed. Fernanda Gil Lozano, Valeria Pita, and Gabriela Ini (Buenos Aires: Taurus, 2000), 111–29; Caimari, "Whose Prisoners Are These?" 185–208; Donna Guy, *Sex and Danger in Buenos Aires: Prostitution, Family, and Nation in Argentina* (Lincoln: University of Nebraska Press, 1990); Kristin Ruggiero, "Honor, Maternity, and the Disciplining of Women: Infanticide in Late Nineteenth-Century Buenos Aires," *Hispanic American Historical Review* 72, no. 3 (1992): 353–72.

25. Julia Rodriguez, "The Argentine Hysteric: A Turn-of-the-Century Psychiatric Type," in *Argentina on the Couch: Psychiatry, State, and Society, 1880 to the Present*, ed. Mariano Plotkin (Albuquerque: University of New Mexico Press, 2003), 25–48.

26. Amador L. Lucero and Lucio V. López, "Melancolía é incapacidad civil," *APCCA* 9 (1910): 180–83; Lucio V. López and Amador L. Lucero, "Demencia inicial é incapacidad civil," *APCCA* 9 (1910): 92–95.

27. Lucero and López, "Melancolía é incapacidad civil," 182–83.

28. José Ingenieros, *Histeria y sugestión. Ensayos de psicología clínica* (1904; 5th ed., Buenos Aires: L. J. Rosso, 1919), 81–82, 8–9.

29. Asunción Lavrin, *Women, Feminism, and Social Change in Argentina, Chile, and Uruguay, 1890–1940* (Lincoln: University of Nebraska Press, 1995); Guy, *Sex and Danger in Buenos Aires.*

30. Bernardo Etchepare, "Puerilismo mental," *APCCA* 6 (1907): 327.

31. Francisco Netri, "El histerismo en la criminalidad," *APCCA* 1 (1902): 145. Emphasis in original.

32. "Las vísperas electorales," *La Nación*, 4 April 1912; "Ecos del día. La elección," *La Nación*, 7 April 1912.

33. Full citizenship was universally denied to women until well into the twentieth century in all liberal republics. See Ruth Lister, *Citizenship: Feminist Perspectives* (New York: New York University Press, 1997); on the United States, see Linda K. Kerber, *No Constitutional Right to Be Ladies: Women and the Obligations of Citizenship* (New York: Hill and Wang, 1998).

34. Hilda Sábato, *The Many and the Few: Political Participation in Republican Buenos Aires* (Stanford, Calif.: Stanford University Press, 2001).

35. "El voto y la acción política," *La Vanguardia*, 8 March 1912.

36. Translated and cited in Kif Augustine-Adams, " 'She Consents Implicitly': Women's Citizenship, Marriage, and Liberal Political Theory in Late Nineteenth- and Early Twentieth-Century Argentina," *Journal of Women's History* 13, no. 4 (2002): 15.

37. James R. Scobie, *Buenos Aires: Plaza to Suburb, 1870–1910* (New York: Oxford University Press, 1974), 235–38.

38. "Proyecto del Diputado Dr. José Miguel Olmedo, 28 May 1890," in Cayetano Carbonell, *Orden y trabajo*, 2 vols. (Buenos Aires: Lajouane, 1910), 1:232–33.

39. "Proyecto del Diputado Dr. Miguel G. Morel, 17 August 1898," in Carbonell, *Orden y trabajo*, 1:240.

40. Ibid., 1:242.

41. F. A. Barroetaveña, *Naturalización de extranjeros*, (Buenos Aires: M. Biedma é hijo, 1909), 43, 46, 47, 3.

42. Adolfo Dickmann, *Los argentinos naturalizados en la política* (Buenos Aires: Talleres Gráficos Juan Perrotti, 1915), 36, 37.

43. Santiago Vaca-Guzmán, *La naturalización de los extranjeros* (Buenos Aires: Casa Editoria Jacobo Peuser, 1891), 68–69.

44. "Proyecto del Diputado Dr. Lucas Ayarragaray, 1 July 1908," in Carbonell, *Orden y trabajo*, 1:249–52.

45. Carbonell, *Orden y trabajo*, 1:227.

46. Alsina, *La inmigración Europea en la República Argentina*, 331. Emphasis in original.

47. José Gregorio Rossi, "La delicuencia profesional en Buenos Aires," *APCCA* 2 (1903): 173.

48. Eusebio Gómez, "Ladrones y usuarios," chap. 3 of *La mala vida en Buenos Aires* (Buenos Aires: Juan Roldán, 1908).

49. Victor Mercante, "Estudio sobre criminalidad infantil," *APCCA* 4 (1905): 578.

50. Victor Arreguine, "El suicidio," *APCCA* 4 (1905): 706.

51. Carlos Alfredo Becú, "La moral de la lucha por la vida," *APCCA* 2 (1903): 518–19.

52. Cornelio Moyano Gacitúa, *La delincuencia argentina* (Cordoba: F. Domenici, 1905), 19.

53. Alsina, *La inmigración Europea en la República Argentina*, 332.

54. José Ingenieros, "Traité de biologie," *APCCA* 2 (1903): 573.

55. Becú, "La moral de la lucha por la vida," 518.

56. Carlos F. Gómez, *Ciudadanía y naturalización. Proyecto de ley.* (Buenos Aires: Alsina, 1913), 125, 126, 32. Emphasis in original.

57. Carbonell, *Orden y trabajo*, 2:3.

58. *Memoria de la Comisión del Centenario al poder ejecutivo nacional* (Buenos Aires: Coni Hermanos, 1910), 15–16, 69–70.

59. Scobie, *Buenos Aires*, 240.

60. Rubén Darío, "Argentina," translated by Patricia Owen Steiner and cited in *The Argentina Reader: History, Culture, Politics*, ed. Gabriela Nouzeilles and Graciela Montaldo (Durham, N.C.: Duke University Press, 2002), 206–7.

61. Gilberto Ramírez, "The Reform of the Argentine Army, 1890–1904" (Ph.D. diss., University of Texas at Austin, 1987), v, 436–41, 445–60; Leopoldo F. Rodríguez, *Inmigración, nacionalismo, y fuerzas armadas. Antecedentes del golpismo en Argentina (1870–1930)* (Mexico City: Editora e Impresora Internacional, 1986), 62–64, 70, 76–77.

62. Translated and cited in Scobie, *Buenos Aires*, 241–44.

63. Earl T. Glaubert, "Ricardo Rojas and the Emergence of Argentine Cultural Nationalism," *Hispanic American Historical Review* 43, no. 1 (1963): 1–13.

64. Richard Slatta, "The Gaucho in Argentina's Quest for National Identity," *Canadian Review of Studies in Nationalism* 12, no. 1 (1985): 99–123.

65. Carbonell, *Orden y trabajo*, 1:226.

66. Cited and translated in Sandra McGee Deutsch, *Counterrevolution in Argentina, 1900–1932: The Argentine Patriotic League* (Lincoln: University of Nebraska Press, 1986), 42.

Chapter Ten

1. On anarchism in Argentina, see Juan Suriano, *Anarquistas. Cultura política y libertarian en Buenos Aires, 1890–1914* (Buenos Aires: Manantial, 2001).

2. For examples of the coverage of the strikes, see *La Prensa*, 22 November 1902, 5; 23 November 1902, 5–6; and 24 November 1902, 5.

3. Sandra McGee Deutsch, *Counterrevolution in Argentina, 1900–1932: The Argentine Patriotic League* (Lincoln: University of Nebraska Press, 1986), 35–40; Sandra McGee Deutsch, *Las Derechas: The Extreme Right in Argentina, Brazil, and Chile, 1890–1939* (Stanford, Calif.: Stanford University Press, 1999), 26–37.

4. *DSCS*, 27 June 1910, 1:296.

5. *DSCS*, 27 June 1910, 1:328.

6. Enrico Ferri, *The Positive School of Criminology: Three Lectures by Enrico Ferri*, ed. Stanley E. Grupp (Pittsburgh, Pa.: University of Pittsburgh Press, 1968).

7. Virgilio Ducceschi, "La criminología moderna," *APCCA* 12 (1913): 419–20.

8. O. Gonzáles Roura, "Delitos contra la honestidad," APCCA 1 (1902): 659, 662.

9. José Ingenieros, "El delito y la defensa social," *APCCA* 8 (1909): 211.

10. José Ingenieros, *La defensa social* (Buenos Aires: Talleres Gráficos de la Penitenciaría Nacional, 1911), 9, 18.

11. Pedro Dorado, "Un derecho penal sin delito y sin pena," *APCCA* 10 (1911): 57.

12. Francisco del Greco, "La psicopatología sintética y aplicada," *APCCA* 5 (1906): 14.

13. José Ingenieros, *Criminología* (1907; 6th ed., Buenos Aires: L. J. Rossi, 1916), 288.

14. *Proyecto de código de procedimientos . . . (1913)* (Buenos Aires: Talleres Gráficos de la Penitenciaría Nacional, 1913), xlviii.

15. *Proyecto de código penal para la República Argentina, . . . 1904* (Buenos Aires: Tipografía de la Cárcel de Encausados, 1906), xx, lxvi, 91–93.

16. Policía de Buenos Aires, *Memoria, 1900–1909*, cited in Juan Suriano, " 'Eliminar los focos de Patología Social,' " in *Historia Social, 1900–2000*, ed. Enrique Masés (Rio Negro: PubliFadecs, 2000), 234.

17. Francisco de Veyga, "Anarquismo y anarquistas. Estudio de antropología criminal," *ADNH* 7, no. 20 (1897): 437–55.

18. Julia Kirk Blackwelder and Lyman L. Johnson, "Changing Criminal Patterns in Buenos Aires, 1890–1914," *Journal of Latin American Studies* 14, no. 2 (1982): 359–79.

19. Miguel Cané, *Expulsión de extranjeros (proyecto de ley)* (Buenos Aires: J. Sarrailh, 1899), 3.

20. *La Prensa*, 9 June 1899, 4–5, 7.

21. Cané, *Expulsión de extranjeros*, 120–21, 24, 57, 11, 125.

22. *DSCD*, 22 November 1902, 1:345–46.

23. *DSCD*, 27 June 1910, 328–29.

24. *DSCS*, 27 November 1902, 1:360, 415; *DSCS*, 22 November 1902, 1:348.

25. *DSCS*, 22 November 1902, 1:361, 348.

26. Francisco Durá, *Naturalización y expulsión de extranjeros* (Buenos Aires: Coni Hermanos, 1911), 7–8.

27. Ibid., 7–8, 282, 283, 284–85, 291.

28. Juan Suriano, *Trabajadores, anarquismo y Estado represor. De la Ley de Residencia a la Ley de Defense Social (1902–1910)* (Buenos Aires: CEAL, 1988), 2.

29. Lyman L. Johnson, "Preface," in *The Problem of Order in Changing Societies: Essays on Crime and Policing in Argentina and Uruguay, 1750–1940*, ed. Lyman L. Johnson (Albuquerque: University of New Mexico Press, 1990), x.

30. *DSCS*, 27 June 1910, 1:326, 314, 316, 296, 298, 302.

31. Sotero F. Vázquez, "Condena de Radowisky. Autor del doble homicidio Falcón-Lartigau," *APCCA* 9 (1910): 359.

32. *DSCS*, 27 June 1910, 1:316.

33. *DSCS*, 27 June 1910, 1:328, 303, 325, 300, 297, 363, 331.

34. Proyecto del diputado doctor Lucas Ayarragaray, Cámara de Diputados, sesión de 10. de julio de 1908, cited in Durá, *Naturalización y expulsión de extranjeros*, 335.

35. *DSCS*, 27 June 1910, 1:301.

36. *DSCS*, 27 June 1910, 1:295, 316, 303, 298, 315.

37. *DSCS*, 22 November 1902, 1:352.

38. *DSCD*, 27 July 1910, 1:359–61.

39. "El estado de sitio," *La Vanguardia*, 5 October 1905.

40. "La Ley de Residencia," *La Vanguardia*, 24 February 1909.

41. "Los extranjeros peligrosos en Córdoba," *La Voz del Pueblo*, 7 February 1903. See also "Nuestras leyes en España," *La Voz del Pueblo*, 24 January 1903.

42. "En plena tiranía," *Vida Nueva*, 19 December 1903.

43. *Proyecto de código de procedimientos . . . (1913)*, lxxvi.

44. O. González Roura, "Es conveniente la reforma proyectada en la legislación penal de fondo?" *RCPML* 1 (1914): 289.

45. *DSCS*, 22 November 1902, 1:361.

46. *Memoria del departamento de Policía de la Capital, 1883–1884* (Buenos Aires: La Tribuna Nacional, 1884), 81, 82.

47. "Expulsión de extranjeros," *RP* 4, no. 51 (1 July 1899): 35–38, 53–56.

48. *Memoria del departamento de la Policía de la Capital, 1906–1909* (Buenos Aires: Imprenta y Encuadernación de la Policía, 1909), 15–16.

49. Durá, *Naturalización y expulsión de extranjeros*, 280.

50. *Memoria del departamento de la Policía de la Capital, 1911–1912* (Buenos Aires: Imprenta y Encuadernación de la Policía, 1912), 5.

51. Ibid., 18.

52. *Memoria del departamento de la Policía de la Capital, 1906–1909*, 298.

53. See Julia Rodriguez, "South Atlantic Crossings: Fingerprints, Science, and the State in Turn-of-the-Century Argentina," *American Historical Review* 109, no. 2 (2004): 387–416.

54. Marcos A. Figueroa, "Las impresiones digitales como prueba de la identidad en los procesos criminales," *APCCA* 12 (1913): 157.

55. Luis Reyna Almandos, *Dactiloscopia argentina. Su historia y influencia en la legislación* (1909; La Plata: University of La Plata, 1932), 109.

56. Juan Vucetich, "Historia sintética de la identificación," *Revista de Identificación y Ciencas Penales* 7 (1931): 91.

57. Ibid., 91.

58. Nicéforo Castellano, *Oficina Nacional de Identidad. Proyecto presentado al Ministerio del Interior* (Buenos Aires: Esuela Tipográfica del Colegio Pio IX, 1908).

59. Cited in Vucetich, "Historia sintética," 101.

60. Vicente Centurion, "La profilaxia de la sifilis," *APCCA* 8 (1909): 301–6.

61. Reyna Almandos, *Dactiloscopia argentina*, 123, 126–27.

62. César Etcheverry to Juan Vucetich, 11 February 1909, correspondence folder, AV.

63. See *Diario de Sesiones de la Cámara de Diputados de la Provincia de Buenos Aires*, 22 September 1909.

64. George Wilton, *Fingerprints: History, Law, and Romance* (London: William Hode, 1938), 86.

65. Cited in Juan Vucetich, *Proyecto de Ley de Registro General de Identificación por Juan Vucetich. Prólogo, notas y apéndice adicional* (La Plata: Taller de Impresiones Oficiales, 1929), 129.

66. Jorge A. Susini to numerous recipients, March 1914, file "Instituto Dactiloscopico Argentino," AV.

67. Luis Reyna Almandos to numerous recipients, March 1914, file "Instituto Dactiloscopico Argentino," AV.

68. Letter to provincial delegates, March 1914, file "Instituto Dactiloscopico Argentino," AV.

69. Vucetich, *Proyecto de Ley de Registro General de Identificación*, 127.

70. Ibid., 131.

71. Ibid., 131.

72. Ministerio de Gobierno de la Provincia de Buenos Aires, *Registro General de Identificación* (La Plata: Taller de Impresiones Oficiales, 1916), 15.

73. Juan Vucetich, *Comentario al proyecto de ley creando el Registro General de Identificación* (La Plata: Taller de Impresiones Oficiales, 1916), 60.

74. Sislán Rodriguez, *La identificación humana. Historia, sistemas y legislación* (La Plata: Taller de Impresiones Oficiales, 1944), 401.

75. Vucetich, *Proyecto de Ley de Registro General de Identificación*, 31, 55.

76. Ministerio de Gobierno de la Provincia de Buenos Aires, *Registro General de Identificación*, 8–9.

77. Kristin Ruggiero, "Fingerprinting and the Argentine Plan for Universal Identification in the Late Nineteenth and Early Twentieth Centuries," in *Documenting Individual Identity: The Development of State Practices in the Modern World*, ed. Jane Caplan and John Torpey (Princeton, N.J.: Princeton University Press, 2001), 185–88.

78. *DSCS*, 27 June 1910, 1:327.

79. Victor Arreguine, "Latinos y anglosajones," *APCCA* 5 (1906): 181–200; José Ingenieros, "Las razas inferiores," in *Crónicas de viaje (1905–06)* (Buenos Aires: L. J. Rosso, 1919), 161–72.

80. Frederick Boyle, "El color rubio," *APCCA* 11 (1912): 126.

81. Nancy Leys Stepan, *"The Hour of Eugenics": Race, Gender, and Nation in Latin America* (Ithaca, N.Y.: Cornell University Press, 1991), 59–60, 82, chap. 4.

82. Natalia Milanesio, "Redefining Men's Sexuality, Re-signifying Male Bodies: The Argentine Law of Anti-Venereal Disease, 1936," *Gender and History* 71, no. 1 (2005).

83. Jorge Salessi, *Médicos maleantes y maricas. Higiene, criminología y homosexualidad en la construcción de la nación Argentina (Buenos Aires: 1871–1914)* (Buenos Aires: Beatríz Viterbo Editora, 1995), 351–59, 372–76.

84. Carlos F. Roche, "El pseudo-hermafrodismo masculino y los androginoides," *APCCA* 3 (1904): 423, 430, 445.

85. Eusebio Gómez, *La mala vida en Buenos Aires* (Buenos Aires: Juan Roldán, 1908), 181.

86. Cited and translated in Eduardo A. Zimmermann, "Racial Ideas and Social Reform: Argentina, 1890–1916," *Hispanic American Historical Review* 72, no. 1 (1992): 42, 44.

87. Ibid., 42.

88. Arturo R. Rossi, president of the Association of Biotypology, Eugenics, and Social Medicine, in 1941; cited and translated in Stepan, *"Hour of Eugenics,"* 141.

Afterword

1. José Ingenieros, *Antología. Su pensamiento en sus majores páginas* (Buenos Aires: Losada, 1961), 25.

2. *Tercer censo nacional de la República Argentina*, 10 vols. (Buenos Aires: Rosso y Compañía, 1916), 1:73.

3. Carlos F. Gómez, *Ciudadanía y naturalización. Proyecto de ley* (Buenos Aires: Alsina, 1913), 23–27.

4. James R. Scobie, *Buenos Aires: Plaza to Suburb, 1870–1910* (New York: Oxford University Press, 1974), 11, 136.

5. José M. Monner Sans, "La function social de nuestra generación," *RCPML* 2 (1915): 294.

6. Miguel A. Lancelotti, "La criminalidad en Buenos Aires, 1887 a 1912," *RCPML* 1 (1914): 20.

7. Luis M. Doyenhard, *La policía en Sud-América* (La Plata: La Popular, 1905), part 1; Roca quoted in David Rock, *Argentina, 1516–1987* (Berkeley: University of California Press, 1987), 155.

8. Luis Alberto Romero, *A History of Argentina in the Twentieth Century* (University Park: Pennsylvania State University Press, 2002), 11; Rock, *Argentina*, 187, 195.

9. Ricardo Salvatore, "Stature, Nutrition, and Regional Convergence: The Argentine Northwest in the First Half of the Twentieth Century," *Social Science History* 28, no. 2 (2004): 297–324.

10. Between 1880 and 1930, the military carried out seventy-three interventions in provincial affairs. See Leopoldo F. Rodríguez, *Inmigración, nacionalismo, y fuerzas armadas. Antecedentes del golpismo en Argentina, 1870–1930* (Mexico City: Editora e Impresora Internacional, 1986), 97, 466–71. See also Gilberto Ramírez, "The Reform of the Argentine Army, 1890–1904" (Ph.D. diss., University of Texas at Austin, 1987), v.

11. See David Rock, *Politics in Argentina, 1890–1930: The Rise and Fall of Radicalism* (Cambridge, U.K.: Cambridge University Press, 1975).

12. Cited and translated in Nancy Leys Stepan, *"The Hour of Eugenics": Race, Gender, and Nation in Latin America* (Ithaca, N.Y.: Cornell University Press, 1991), 141.

13. See Uki Goñi, *The Real Odessa: Smuggling the Nazis to Perón's Argentina* (London: Granta Press, 2002).

14. Carlos Waisman, *The Reversal of Development in Argentina: Postwar Counterrevolutionary Policies and Their Structural Consequences* (Princeton, N.J.: Princeton University Press, 1987), 228.

15. Marguerite Feitlowitz, *A Lexicon of Terror: Argentina and the Legacies of Torture* (New York: Oxford University Press, 1998), 33.

16. Translated and cited in ibid., 7, 33, 44.

17. See Laura Kalmanowiecki, "Policing the People, Building the State: The Police-

Military Nexus in Argentina, 1880–1945," in *Irregular Armed Forces and Their Role in Politics and State Formation*, ed. Diane Davis and Anthony Pereira (Cambridge, U.K.: Cambridge University Press, 2003), 209–31. See also "La policía Bonaerense: Sospechas de la justicia y el Congreso. Caso AMIA," and "La policía Bonaerense: El Asinato de José Luis Cabezas," *Clarín*, 19 November 1997; and "Governments vs. Journalists: Latin News Media, Hard-Pressed," *New York Times*, 5 October 1997.

18. See Dora Barrancos, *La escena iluminada. Ciencias para trabajadores, 1890–1930* (Buenos Aires: Plus Ultra, 1996); and Gregorio Weinberg, *La ciencia y el idea del progreso en América Latina, 1860–1930* (Buenos Aires: Fondo de Cultura Económica, 1998).

19. See David Rock, *State Building and Political Movements in Argentina, 1860–1916* (Stanford, Calif.: Stanford University Press, 2002); and Natalio Botana, *El orden conservador. La política argentina entre 1880 y 1916* (Buenos Aires: Sudamerica, 1994).

20. Argentine historian Tulio Halperín Donghi, in a recent survey of views of Argentina's "reversal of fortune," sees a retreat of the old polarizing ideologies, replaced disturbingly by fatalism and apathy. After the state terror of the 1970s and 1980s, and recurring and extreme economic crises, Argentines have become pessimistic and disillusioned, nearly forgetting the earlier periods of prosperity and promise, he argues. See Tulio Halperín Donghi, "Argentines Ponder the Burden of the Past," in *Colonial Legacies: The Problem of Persistence in Latin American History*, ed. Jeremy Adelman (New York: Routledge, 1999), 151–73.

Essay on Sources

Primary Sources

Civilizing Argentina is rooted in a wide range of archival and printed primary sources in Argentina and the United States. My research began at the Columbia University libraries, which house many books on criminology, sociology, and public health written by Argentine physicians, social scientists, and legislators of the late nineteenth and early twentieth centuries. The Columbia Law Library has a full run of the *Archivos de Psiquiatría, Criminología, y Ciencias Afines*, and the Health Sciences Library at the Columbia medical campus possesses additional medical publications and journals. The New York Public Library provided an unexpected trove of published Argentine sources, including newspapers, scientific journals, pamphlets, and government documents, such as the *Anuario Estadístico de la Ciudad de Buenos Aires*.

Harvard University's libraries, too, offer a wealth of materials on Argentine science and medicine. Widener Library has an extensive collection of Argentine scientific and government journals, and the Harvard Law School Library has numerous Argentine legal journals, published laws, and published legal scholarship. Legal scholarly journals, including the *Revista Jurídica*, *Revista de Policía*, and the *Revista Penitenciaría* are available in many libraries in the United States, in particular the Harvard Law School Library. The U.S. Library of Congress has available full runs of the Argentine legislative record, along with some harder-to-find journals and many excellent photographs of turn-of-the-century Argentina.

In Argentina I used a number of government and university archives, some of which provided more ready access to materials and ease of research than others. The two most hospitable archives are the Archivo General de la Nación (AGN) and the University of Buenos Aires (UBA) medical school library and archive. The AGN provides valuable material, though the majority of its holdings are from the pre-1880 period. Among its important twentieth-century documents are correspondence and official reports, including documents from the Archivo Miguel Cané and the Archivo del Ministerio del Interior. At the Archivo de la Municipalidad de la Ciudad de Buenos Aires, I consulted the series *Salud Pública*, *Gobierno*, and *Cultura* for documentation of municipal government policies on hospitals, policing, and the pathology museum. The extensive photography collection at the AGN generously provided many turn-of-the-century photos.

At the Archivo de la Facultad de Medicina (medical school archive) of the University of Buenos Aires, I consulted budgets and personnel and student files from the turn of the century. The library at the medical school provided a vast number of historical medical publications and medical and scientific journals, such as *Anales del Círculo Médico Argentino*; *Anales del Departamento Nacional de Higiene*; *Archivos de Psiquiatría, Criminología, y Ciencias Afines*; *Revista de Criminología, Psiquiatría, y Medicina Legal*; and *Semana Médica*. Many medical, social science, and legal books by turn-of-the-century authors, such as José María Ramos Mejía's *Las neurosis de los hombres célebres en la historia argentina* (Buenos Aires: M. Biedma, 1878), Eusebio Gómez's *La mala vida en Buenos Aires* (Buenos Aires: Juan Roldán, 1908), and Enrique Feinmann's *La ciencia del niño* (Buenos Aires: Cabaut y Compañía, 1915) are scattered in libraries throughout the Americas; but most of them are available in the Argentine libraries mentioned here, in the New York Public Library, at Harvard, and at Columbia. Though the personal papers of pioneering psychiatrist and criminologist José Ingenieros are held by the family and presently unavailable to researchers, his published work, spanning topics from psychiatry to sociology, is available in collected volumes, including *Obras completas* (Buenos Aires: Mar Océano, 1962). I also consulted dozens of doctoral dissertations, in medicine as well as law, in the Colección Candioti at the Biblioteca Nacional (Argentine national library) in Buenos Aires. A useful index of these dissertations can be found in *Catálogo de la colección de tésis, 1827–1917* (Buenos Aires: A. Flabian, 1918) and in Marcial R. Candioti, *Bibliografía doctoral de la Universidad de Buenos Aires y catálogo cronológico de las tésis en su primer centenario 1821–1920* (Buenos Aires: Ministerio de Agricultura, 1920). The older periodical division at the Biblioteca Nacional holds the only full run, as far as I know, of the rare criminological journal, *Criminalogía Moderna*.

Certain records were more difficult to obtain, especially on police and prison history. The police library Archivo Romay, in Buenos Aires, has relatively open access to its limited collection of documents on the capital police's history, but the police archive, the Archivo de la Policía de la Capital, Buenos Aires, generally does not admit outside researchers. Through serendipity, I did gain access to this archive, and I was able to take notes on the nineteenth-century police ledgers called the *Copiadores de Notas*.

In the city of La Plata, where the Archivo Vucetich at the Museo de la Policía de la Provincia de Buenos Aires was closed for renovations, I acquired permission to examine the uncataloged papers from Juan Vucetich's files, not only on his work but also on the history of the provincial police and the fingerprint office, records that have been largely unconsulted for decades, an unanticipated discovery that significantly shaped this book. At the Archivo Vucetich, I examined these folders (at the time, uncataloged): Correspondencia; Oficina Dactiloscopica; Impresiones Digitales; Oficina Central de Investigación de la Provincia; Museo Vucetich; Notas y folios; Policía Federal; and Centro de Estudios Policiales.

I have used statistics on crime and arrest rates from the three national censuses of the period and in the *Anuario Estadístico de la Ciudad de Buenos Aires*, available at the New York Public Library. Books, such as Alberto Méndez Casariego's *La criminalidad de la ciudad de Buenos Aires en 1887. Contravenciones, suicidios, y accidentes* (Buenos Aires: Imprenta del Departamento de Policía de la Capital, 1888), also contain crime statistics. I consulted legislative debates and records of voting and laws, the *Diario de Sesiones de la Cámara de Senadores* and the *Diario de Sesiones de la Cámara de Diputados*, at the Biblioteca del Congreso (Argentine Library of Congress) in Buenos Aires (these are also available in many U.S. university libraries).

Conducting research on prison history in Argentina proved difficult because not all important data is yet accessible. The Museo Penitenciario Argentino "Antonio Ballvé" (Argentine penitentiary museum) in Buenos Aires permits researchers access to only a few of its holdings. One elusive—and important—source for scholars is the collection of original biopsychological profiles of prisoners in the National Penitentiary for the years before the 1930s. The vast majority of these original documents from the Criminology Institute, which was located in the penitentiary, have apparently been misplaced or destroyed. My discussion of patient cases in *Civilizing Argentina*, therefore, necessarily depends on published observations of inmates reprinted in the journals *Revista Penitenciaría* and *Archivos de Psiquiatría, Criminología, y Ciencias Afines*. I did, however, examine the files called "Prontuarios de presos, Penitenciaría Nacional"; the "Libros de Entradas de Internos, 1899–1908"; the ledger "Contraventoras procedadas y penadas, 1912–1922"; and the "Testamonios de Sentencias, 1897–1907." In La Plata, the Servicio Penitenciario de la Provincia de Buenos Aires provided personal anthropometric descriptions and additional documentation on the movement of prisoners in the folders "Libros de Entradas y Salidas de Presos Criminales"; "Fichas de Filiación"; and "Cárceles de la Provincia."

Because the dynamic between late nineteenth-century scientific thought and state practices in Argentina developed as a result of social interaction among scientists, politicians, and the subjects of their focus and is not reflected in formal, scientific thought alone but in the bureaucracies of police and public health systems and in the press and published literature, I sought to assess the popular reception of the new scientific ideas in daily newspapers such as *La Nación* and *La Prensa*, both filed at the Biblioteca Nacional and available on microfilm in many U.S. libraries. I consulted *Caras y Caretas*, a weekly magazine read by the Argentine middle class, and socialist and anarchist newspapers, including *La Vanguardia* and *La Voz del Pueblo*. While left-wing and feminist papers, many of which were published only for a few months, are difficult to find in their full runs, groups of issues are available through interlibrary loan.

Principal Published and Scholarly Studies

On Argentine "civilization," the most useful studies on the nineteenth-century national founding of Argentina are Nicolas Shumway, *The Invention of Argentina: History of an Idea* (Berkeley: University of California Press, 1991) and Francine Masiello, *Between Civilization and Barbarism: Women, Nation, and Literary Culture in Modern Argentina* (Lincoln: University of Nebraska Press, 1992). An influential sociological examination of twentieth-century sociopolitical patterns, Carlos Waisman's *The Reversal of Development in Argentina: Postwar Counterrevolutionary Policies and Their Structural Consequences* (Princeton, N.J.: Princeton University Press, 1987), addresses the main question of Argentina's seeming decline from prosperity to chaos. On state formation in turn-of-the-century Argentina, useful books are David Rock, *State Building and Political Movements in Argentina, 1860–1916* (Stanford, Calif.: Stanford University Press, 2002); and Natalio Botana, *El orden conservador. La política argentina entre 1880 y 1916* (4th ed., Buenos Aires: Sudamerica, 1994).

Few historical analyses of racial conflict in Argentina exist, with an especially glaring lack of information on Argentines of African origin. The only study in print is George Reid Andrews, *The Afro-Argentines of Buenos Aires, 1800–1900* (Madison: University of Wisconsin Press, 1980). There is a need for more research on the multiracial nature of Argentine society; on this omission in Argentine history, see Susana Rotker, *Captive Women: Oblivion and Memory in Argentina* (Minneapolis: University of Minnesota Press, 2002), especially chapter 2.

Late nineteenth-century discourse on Argentina's immigration was rife with racial meanings. Among the most useful studies on this theme are José C. Moya, *Cousins and Strangers: Spanish Immigrants in Buenos Aires, 1850–1930* (Berkeley: University of California Press, 1998), especially chapter 2; on Italian immigration to Argentina in comparative perspective, see Samuel L. Baily, "The Adjustment of Italian Immigrants in Buenos Aires and New York, 1870–1914," *American Historical Review* 88, no. 2 (1983): 281–305, and Herbert S. Klein, "AHR Forum: The Integration of Italian Immigrants into the United States and Argentina: A Comparative Analysis," *American Historical Review* 88, no. 2 (1983): 306–46. On xenophobia in Argentina, see Carl Solberg, *Immigration and Nationalism: Argentina and Chile, 1890–1914* (Austin: University of Texas Press, 1970). On the anarchists, see Juan Suriano, *Trabajadores, anarquismo y Estado represor. De la Ley de Residencia a la Ley de Defensa Social (1902–1910)* (Buenos Aires: CEAL, 1988).

A number of recent studies by intellectual and medical historians of Argentina have argued convincingly that traditional political and economic histories are not sufficient to understand the changing language, methods, and concepts behind elite attempts at social control, and that medical and scientific ideas themselves inspired new social practices and carried great weight in the attempt to harness the unpredictable effects of progress. Those sources I found most useful were Nancy Leys Stepan, *"The Hour of Eugenics": Race, Gender, and Nation in Latin America*

(Ithaca, N.Y.: Cornell University Press, 1991); Oscar Terán, *Vida intellectual en el Buenos Aires fin-de-siglo (1880–1910). Derivas de la "cultura científica"* (Buenos Aires: Fondo de Cultura Económica, 2000); Eduardo Zimmermann, *Los liberales reformistas. La cuestión social en la Argentina, 1890–1916* (Buenos Aires: Editorial Sudamericana, 1994); and Donna Guy, *Sex and Danger in Buenos Aires: Prostitution, Family, and Nation in Argentina* (Lincoln: University of Nebraska Press, 1990).

The social impact of medicine in Argentina is a topic just now being explored and awaits further studies. On the history of sanitation and public health in turn-of-the-century Argentina, see Jorge Salessi, *Médicos maleantes y maricas: Higiene, criminología y homosexualidad en la construcción de la nación Argentina (Buenos Aires: 1871–1914)* (Buenos Aires: Beatríz Viterbo Editora, 1995). An initial study of the ties between medicine and law can be found in Mirta Zaida Lobato, ed., *Política, médicos y enfermedades. Lecturas de historia de la salud en la Argentina* (Buenos Aires: Editorial Biblos, 1996). However, further research will be necessary in order to understand more fully the practical outcomes of medicine in public and private spheres, including the role of women physician-reformers and their impact on legislation and policy. For the history of psychology in Argentina, see Mariano Plotkin, *Freud in the Pampas: The Emergence and Development of a Psychoanalytic Culture in Argentina* (Stanford, Calif.: Stanford University Press, 2001); and Mariano Plotkin, ed., *Argentina on the Couch: Psychiatry, State, and Society, 1880 to the Present* (Albuquerque: University of New Mexico Press, 2003).

On nineteenth-century growth of the Buenos Aires police, see Osvaldo Barre-neche, *Dentro de la ley, todo: La justicia criminal en Buenos Aires en la etapa formativa del sistema penal moderna de la Argentina* (La Plata: Ediciones al margen, 2001). On political and social policing, see Laura Kalmanowiecki, "Policing the People, Building the State: The Police-Military Nexus in Argentina, 1880–1945," in *Irregular Armed Forces and Their Role in Politics and State Formation*, ed. Diane Davis and Anthony Pereira (Cambridge, U.K.: Cambridge University Press, 2003), 209–31. On fingerprinting as a means of social control seen in global context, see Julia Rodriguez, "South Atlantic Crossings: Fingerprints, Science, and the State in Turn-of-the-Century Argentina," *American Historical Review* 109, no. 2 (2004): 387–416.

Two recent books that analyze outcomes and cultural meanings in prisons and criminal court trials, respectively, are Lila M. Caimari, *Apenas un delinceunte. Crimen, castigo y cultura en Buenos Aires (1877–1955)* (Buenos Aires: Siglo XXI, 2004); and Kristin Ruggiero, *Modernity in the Flesh: Medicine, Law, and Society in Turn-of-the-Century Argentina* (Stanford, Calif.: Stanford University Press, 2004). Criminal justice institutions in the post-1930 period remain virtually uninvestigated; future studies should take up the role of scientific criminology and policing under the military dictatorships.

On the importance of labor and work discipline to the prison reform efforts at the turn of the century, see Ricardo D. Salvatore, "Criminology, Prison Reform, and the Buenos Aires Working Class," *Journal of Interdisciplinary History* 23, no. 2 (1992): 279–99. For an interesting analysis of prisoner examinations conducted at the

Instituto de Criminología in a later period, see Lila M. Caimari, "Remembering Freedom: Life as Seen from the Prison Cell," in *Crime and Punishment in Latin America: Law and Society Since Late Colonial Times*, ed. Ricardo Salvatore, Carlos Aguirre, and Gilbert Joseph (Durham, N.C.: Duke University Press, 2001), 391–414. On women in confinement, see Lila M. Caimari, "Whose Criminals Are These? Church, State, and *Patronatos* and the Rehabilitation of Female Criminals (Buenos Aires, 1890–1970)," *Americas* 54, no. 2 (1997): 185–208; Donna Guy, "Girls in Prison: The Role of the Buenos Aires Casa Correccional de Mujeres as an Institution for Child Rescue, 1890–1940," also in *Crime and Punishment in Latin America*, 361–90. On houses of deposit, see Kristin Ruggiero, "Houses of Deposit and the Exclusion of Women in Turn-of-the-Century Argentina," in *Isolation: Places and Practices of Exclusion*, ed. Carolyn Strange and Alison Bashford (New York: Routledge, 2003), 119–32. On mental institutions, see Jonathan Ablard, "Law, Medicine, and Confinement to Public Psychiatric Hospitals in Twentieth-Century Argentina," in *Argentina on the Couch: Psychiatry, State, and Society, 1880 to the Present*, ed. Mariano Plotkin (Albuquerque: University of New Mexico Press, 2003), 87–112.

There are a number of histories of the socialist, anarchist, and workers' mobilizations in Argentina; the most useful is Juan Suriano, *Anarquistas. Cultura política y libertarian en Buenos Aires, 1890–1914* (Buenos Aires: Manantial, 2001). Less studied is the struggle for women's legal rights; see Asunción Lavrin, *Women, Feminism, and Social Change in Argentina, Chile, and Uruguay, 1890–1940* (Lincoln: University of Nebraska Press, 1995). On women and citizenship rights, see Kif Augustine-Adams, " 'She Consents Implicitly': Women's Citizenship, Marriage, and Liberal Political Theory in Late Nineteenth- and Early Twentieth-Century Argentina," *Journal of Women's History* 13, no. 4 (2002): 8–30. Further research on legislative change, including the criminal and civil codes, and their social impact on men and women alike, awaits attention by scholars.

Index